D1068597

Pitman Research Notes in Mathematics Series

Submission of proposals for consideration
Suggestions for publication, in the form of outlines and representative samples, are invited by the Editorial Board for assessment. Intending authors should approach one of the main editors or another member of the Editorial Board, citing the relevant AMS subject classifications. Alternatively, outlines may be sent directly to the publisher's offices. Refereeing is by members of the board and other mathematical authorities in the topic concerned, throughout the world.

Preparation of accepted manuscripts
On acceptance of a proposal, the publisher will supply full instructions for the preparation of manuscripts in a form suitable for direct photo-lithographic reproduction. Specially printed grid sheets are provided and a contribution is offered by the publisher towards the cost of typing. Word processor output, subject to the publisher's approval, is also acceptable.

Illustrations should be prepared by the authors, ready for direct reproduction without further improvement. The use of hand-drawn symbols should be avoided wherever possible, in order to maintain maximum clarity of the text.

The publisher will be pleased to give any guidance necessary during the preparation of a typescript, and will be happy to answer any queries.

Important note
In order to avoid later retyping, intending authors are strongly urged not to begin final preparation of a typescript before receiving the publisher's guidelines and special paper. In this way it is hoped to preserve the uniform appearance of the series.

Longman Scientific & Technical
Longman House
Burnt Mill
Harlow, Essex, UK
(tel (0279) 426721)

Titles in this series

1 Improperly posed boundary value problems
A Carasso and A P Stone

2 Lie algebras generated by finite dimensional ideals
I N Stewart

3 Bifurcation problems in nonlinear elasticity
R W Dickey

4 Partial differential equations in the complex domain
D L Colton

5 Quasilinear hyperbolic systems and waves
A Jeffrey

6 Solution of boundary value problems by the method of integral operators
D L Colton

7 Taylor expansions and catastrophes
T Poston and I N Stewart

8 Function theoretic methods in differential equations
R P Gilbert and R J Weinacht

9 Differential topology with a view to applications
D R J Chillingworth

10 Characteristic classes of foliations
H V Pittie

11 Stochastic integration and generalized martingales
A U Kussmaul

12 Zeta-functions: An introduction to algebraic geometry
A D Thomas

13 Explicit *a priori* inequalities with applications to boundary value problems
V G Sigillito

14 Nonlinear diffusion
W E Fitzgibbon III and H F Walker

15 Unsolved problems concerning lattice points
J Hammer

16 Edge-colourings of graphs
S Fiorini and R J Wilson

17 Nonlinear analysis and mechanics: Heriot-Watt Symposium Volume I
R J Knops

18 Actions of fine abelian groups
C Kosniowski

19 Closed graph theorems and webbed spaces
M De Wilde

20 Singular perturbation techniques applied to integro-differential equations
H Grabmüller

21 Retarded functional differential equations: A global point of view
S E A Mohammed

22 Multiparameter spectral theory in Hilbert space
B D Sleeman

24 Mathematical modelling techniques
R Aris

25 Singular points of smooth mappings
C G Gibson

26 Nonlinear evolution equations solvable by the spectral transform
F Calogero

27 Nonlinear analysis and mechanics: Heriot-Watt Symposium Volume II
R J Knops

28 Constructive functional analysis
D S Bridges

29 Elongational flows: Aspects of the behavi
of model elasticoviscous fluids
C J S Petrie

30 Nonlinear analysis and mechanics: Heriot-Watt Symposium Volume III
R J Knops

31 Fractional calculus and integral transforms
generalized functions
A C McBride

32 Complex manifold techniques in theoretic
physics
D E Lerner and P D Sommers

33 Hilbert's third problem: scissors congruen
C-H Sah

34 Graph theory and combinatorics
R J Wilson

35 The Tricomi equation with applications to theory of plane transonic flow
A R Manwell

36 Abstract differential equations
S D Zaidman

37 Advances in twistor theory
L P Hughston and R S Ward

38 Operator theory and functional analysis
I Erdelyi

39 Nonlinear analysis and mechanics: Heriot-Watt Symposium Volume IV
R J Knops

40 Singular systems of differential equations
S L Campbell

41 N-dimensional crystallography
R L E Schwarzenberger

42 Nonlinear partial differential equations in physical problems
D Graffi

43 Shifts and periodicity for right invertible operators
D Przeworska-Rolewicz

44 Rings with chain conditions
A W Chatters and C R Hajarnavis

45 Moduli, deformations and classifications of compact complex manifolds
D Sundararaman

46 Nonlinear problems of analysis in geometry and mechanics
M Atteia, D Bancel and I Gumowski

47 Algorithmic methods in optimal control
W A Gruver and E Sachs

48 Abstract Cauchy problems and functional differential equations
F Kappel and W Schappacher

49 Sequence spaces
W H Ruckle

50 Recent contributions to nonlinear partial differential equations
H Berestycki and H Brezis

51 Subnormal operators
J B Conway

52 Wave propagation in viscoelastic media
F Mainardi
53 Nonlinear partial differential equations and their applications: Collège de France Seminar. Volume I
H Brezis and J L Lions
54 Geometry of Coxeter groups
H Hiller
55 Cusps of Gauss mappings
T Banchoff, T Gaffney and C McCrory
56 An approach to algebraic K-theory
A J Berrick
57 Convex analysis and optimization
J-P Aubin and R B Vintner
58 Convex analysis with applications in the differentiation of convex functions
J R Giles
59 Weak and variational methods for moving boundary problems
C M Elliott and J R Ockendon
60 Nonlinear partial differential equations and their applications: Collège de France Seminar. Volume II
H Brezis and J L Lions
61 Singular systems of differential equations II
S L Campbell
62 Rates of convergence in the central limit theorem
Peter Hall
63 Solution of differential equations by means of one-parameter groups
J M Hill
64 Hankel operators on Hilbert space
S C Power
65 Schrödinger-type operators with continuous spectra
M S P Eastham and H Kalf
66 Recent applications of generalized inverses
S L Campbell
67 Riesz and Fredholm theory in Banach algebra
B A Barnes, G J Murphy, M R F Smyth and T T West
68 Evolution equations and their applications
F Kappel and W Schappacher
69 Generalized solutions of Hamilton-Jacobi equations
P L Lions
70 Nonlinear partial differential equations and their applications: Collège de France Seminar. Volume III
H Brezis and J L Lions
71 Spectral theory and wave operators for the Schrödinger equation
A M Berthier
72 Approximation of Hilbert space operators I
D A Herrero
73 Vector valued Nevanlinna Theory
H J W Ziegler
74 Instability, nonexistence and weighted energy methods in fluid dynamics and related theories
B Straughan
75 Local bifurcation and symmetry
A Vanderbauwhede
76 Clifford analysis
F Brackx, R Delanghe and F Sommen
77 Nonlinear equivalence, reduction of PDEs to ODEs and fast convergent numerical methods
E E Rosinger
78 Free boundary problems, theory and applications. Volume I
A Fasano and M Primicerio
79 Free boundary problems, theory and applications. Volume II
A Fasano and M Primicerio
80 Symplectic geometry
A Crumeyrolle and J Grifone
81 An algorithmic analysis of a communication model with retransmission of flawed messages
D M Lucantoni
82 Geometric games and their applications
W H Ruckle
83 Additive groups of rings
S Feigelstock
84 Nonlinear partial differential equations and their applications: Collège de France Seminar. Volume IV
H Brezis and J L Lions
85 Multiplicative functionals on topological algebras
T Husain
86 Hamilton-Jacobi equations in Hilbert spaces
V Barbu and G Da Prato
87 Harmonic maps with symmetry, harmonic morphisms and deformations of metrics
P Baird
88 Similarity solutions of nonlinear partial differential equations
L Dresner
89 Contributions to nonlinear partial differential equations
C Bardos, A Damlamian, J I Díaz and J Hernández
90 Banach and Hilbert spaces of vector-valued functions
J Burbea and P Masani
91 Control and observation of neutral systems
D Salamon
92 Banach bundles, Banach modules and automorphisms of C*-algebras
M J Dupré and R M Gillette
93 Nonlinear partial differential equations and their applications: Collège de France Seminar. Volume V
H Brezis and J L Lions
94 Computer algebra in applied mathematics: an introduction to MACSYMA
R H Rand
95 Advances in nonlinear waves. Volume I
L Debnath
96 FC-groups
M J Tomkinson
97 Topics in relaxation and ellipsoidal methods
M Akgül
98 Analogue of the group algebra for topological semigroups
H Dzinotyiweyi
99 Stochastic functional differential equations
S E A Mohammed

100 Optimal control of variational inequalities
V Barbu
101 Partial differential equations and dynamical systems
W E Fitzgibbon III
102 Approximation of Hilbert space operators. Volume II
C Apostol, L A Fialkow, D A Herrero and D Voiculescu
103 Nondiscrete induction and iterative processes
V Ptak and F-A Potra
104 Analytic functions – growth aspects
O P Juneja and G P Kapoor
105 Theory of Tikhonov regularization for Fredholm equations of the first kind
C W Groetsch
106 Nonlinear partial differential equations and free boundaries. Volume I
J I Díaz
107 Tight and taut immersions of manifolds
T E Cecil and P J Ryan
108 A layering method for viscous, incompressible L_p flows occupying R^n
A Douglis and E B Fabes
109 Nonlinear partial differential equations and their applications: Collège de France Seminar. Volume VI
H Brezis and J L Lions
110 Finite generalized quadrangles
S E Payne and J A Thas
111 Advances in nonlinear waves. Volume II
L Debnath
112 Topics in several complex variables
E Ramírez de Arellano and D Sundararaman
113 Differential equations, flow invariance and applications
N H Pavel
114 Geometrical combinatorics
F C Holroyd and R J Wilson
115 Generators of strongly continuous semigroups
J A van Casteren
116 Growth of algebras and Gelfand–Kirillov dimension
G R Krause and T H Lenagan
117 Theory of bases and cones
P K Kamthan and M Gupta
118 Linear groups and permutations
A R Camina and E A Whelan
119 General Wiener–Hopf factorization methods
F-O Speck
120 Free boundary problems: applications and theory, Volume III
A Bossavit, A Damlamian and M Fremond
121 Free boundary problems: applications and theory, Volume IV
A Bossavit, A Damlamian and M Fremond
122 Nonlinear partial differential equations and their applications: Collège de France Seminar. Volume VII
H Brezis and J L Lions
123 Geometric methods in operator algebras
H Araki and E G Effros
124 Infinite dimensional analysis–stochastic processes
S Albeverio
125 Ennio de Giorgi Colloquium
P Krée
126 Almost-periodic functions in abstract spac
S Zaidman
127 Nonlinear variational problems
A Marino, L Modica, S Spagnolo and M Degiovanni
128 Second-order systems of partial differenti equations in the plane
L K Hua, W Lin and C-Q Wu
129 Asymptotics of high-order ordinary differ equations
R B Paris and A D Wood
130 Stochastic differential equations
R Wu
131 Differential geometry
L A Cordero
132 Nonlinear differential equations
J K Hale and P Martinez-Amores
133 Approximation theory and applications
S P Singh
134 Near-rings and their links with groups
J D P Meldrum
135 Estimating eigenvalues with *a posteriori/a* inequalities
J R Kuttler and V G Sigillito
136 Regular semigroups as extensions
F J Pastijn and M Petrich
137 Representations of rank one Lie groups
D H Collingwood
138 Fractional calculus
G F Roach and A C McBride
139 Hamilton's principle in continuum mechanics
A Bedford
140 Numerical analysis
D F Griffiths and G A Watson
141 Semigroups, theory and applications. Volu
H Brezis, M G Crandall and F Kappel
142 Distribution theorems of L-functions
D Joyner
143 Recent developments in structured contin
D De Kee and P Kaloni
144 Functional analysis and two-point differen operators
J Locker
145 Numerical methods for partial differential equations
S I Hariharan and T H Moulden
146 Completely bounded maps and dilations
V I Paulsen
147 Harmonic analysis on the Heisenberg nilpc Lie group
W Schempp
148 Contributions to modern calculus of variati
L Cesari
149 Nonlinear parabolic equations: qualitative properties of solutions
L Boccardo and A Tesei
150 From local times to global geometry, contr physics
K D Elworthy

200 Stochastic analysis, path integration and dynamics
D Elworthy
201 Riemannian geometry and holonomy groups
S Salamon
202 Strong asymptotics for extremal errors and polynomials associated with Erdös type weights
D S Lubinsky
203 Optimal control of diffusion processes
V S Borkar
204 Rings, modules and radicals
B J Gardner
205 Numerical studies for nonlinear Schrödinger equations
B M Herbst and J A C Weideman
206 Distributions and analytic functions
R D Carmichael and D Mitrović
207 Semicontinuity, relaxation and integral representation in the calculus of variations
G Buttazzo
208 Recent advances in nonlinear elliptic and parabolic problems
P Bénilan, M Chipot, L Evans and M Pierre
209 Model completions, ring representations and the topology of the Pierce sheaf
A Carson
210 Retarded dynamical systems
G Stepan
211 Function spaces, differential operators and nonlinear analysis
L Paivarinta
212 Analytic function theory of one complex variable
C C Yang, Y Komatu and K Niino
213 Elements of stability of visco-elastic fluids
J Dunwoody
214 Jordan decompositions of generalised vector measures
K D Schmidt
215 A mathematical analysis of bending of plates with transverse shear deformation
C Constanda
216 Ordinary and partial differential equations Vol II
B D Sleeman and R J Jarvis
217 Hilbert modules over function algebras
R G Douglas and V I Paulsen
218 Graph colourings
R Wilson and R Nelson
219 Hardy-type inequalities
A Kufner and B Opic
220 Nonlinear partial differential equations and their applications. College de France Seminar Volume X
H Brezis and J L Lions
221 Workshop on dynamical systems
E Shiels and Z Coelho
222 Geometry and analysis in nonlinear dynamics
H W Broer and F Takens
223 Fluid dynamical aspects of combustion theory
M Onofri and A Tesei
224 Approximation of Hilbert space operators. Volume I. 2nd edition
D Herrero
225 Operator Theory: Proceedings of the 198
GPOTS–Wabash conference
J B Conway and B B Morrel
226 Local cohomology and localization
J L Bueso Montero, B Torrecillas Jover ar A Verschoren
227 Sobolev spaces of holomorphic functions
F Beatrous and J Burbea
228 Numerical analysis. Volume III
D F Griffiths and G A Watson
229 Recent developments in structured contin Volume III
D De Kee and P Kaloni
230 Boolean methods in interpolation and approximation
F J Delvos and W Schempp
231 Further advances in twistor theory, Volun
L J Mason and L P Hughston
232 Further advances in twistor theory, Volun
L J Mason and L P Hughston
233 Geometry in the neighborhood of invarian manifolds of maps and flows and lineariza
U Kirchgraber and K Palmer
234 Quantales and their applications
K I Rosenthal
235 Integral equations and inverse problems
R Lazarov and V Petkov
236 Pseudo-differential operators
S R Simanca
237 A functional analytic approach to statistica experiments
I M Bomze
238 Quantum mechanics, algebras and distribu
D Dubin and M Hennings

Hamiltonian Flows and Evolution Semigroups

Henryk Gzyl

Universidad Central de Venezuela Caracas, Venezuela

Hamiltonian Flows and Evolution Semigroups

Copublished in the United States with
John Wiley & Sons, Inc., New York

Longman Scientific & Technical,
Longman Group UK Limited,
Longman House, Burnt Mill, Harlow
Essex CM20 2JE, England
and Associated Companies throughout the world.

Copublished in the United States with
John Wiley & Sons, Inc., 605 Third Avenue, New York, NY 10158

First published 1990

AMS Subject Classification: 70H15, 70H99, 47D07, 42A38, 20M99

ISSN 0269-3674

British Library Cataloguing in Publication Data
Gzyl, Henryk
Hamiltonian flows and evolution semigroups.
1. Statistical mechanics
I. Title
530.13

ISBN 0-582-03190-7

Library of Congress Cataloging-in-Publication Data
Gzyl, Henryk, 1946–
Hamiltonian flows and evolution semigroups / Henryk Gzyl.
p. cm.—(Pitman research notes in mathematics series; 239)
Includes bibliographical references.
1. Probabilities. 2. Semigroups. 3. Hamiltonian systems.
I. Title. II. Series.
QA273.G98 1990
512′.2--dc20

90-42155 CIP

Printed and bound in Great Britain
by Biddles Ltd, Guildford and King's Lynn

Contents

Chapter VI Applications 6

reface

eing curious about the connection between polynomials of binomial type and exponentials I ad during 1978 Philip Feinsilver's "Special functions, probability semigroups and amiltonian mechanics". It was hard not to like the interplay of ideas from different fields. lowever, I found two things missing: First, the extension of his scheme to Hamiltonian ystems beyond the trivially integrable ones and second, the theory of canonical trans- rmations was not made use of.

During 1983, while on sabbatical in that interesting and beautiful country, Holland, at what now C.W.I. in Amsterdam, Michiel Hazewinkel suggested my looking at some possible xtensions of the linear filtering problem which I still have not looked at, but that put me in the ght track to continue with Feinsilver's line of work.

The idea for this book was planted in my mind by one of the editors of Addison-Wesley wish I knew who had planted it in his). Anyway, what came out was a potpourri of ideas, e core of which is a theory of non-unitary representations of canonical transformations and ome operational calculus naturally associated with it.

A warning: Do not insist on rigour. There is little, if any, of it in here. However I tried to e precise enough so that statements could be either disproved or their exact validity scertained. Any attempt to make the material in this book rigorous would end up making it o lengthy.

The final form of this book took shape in Carbondale, in a visit organized by Phil einsilver. I owe to him a number of criticisms on the first draft of this material. The present ı. s. is a real improvement on the previous ones.

I should end acknowledging the following institutions, each of which contributed to aking my visit to Carbondale possible: Fundación Polar, CONICIT and C.D.C.H.-U.C.V. n Caracas and the Mathematics Dept. of S.I.U. at Carbondale.

Caracas, 1988.

Chapter I
Introduction

What is this book about?. As mentioned in the preface, this is an attempt to extend the ideas in Feinsilver's work [IV-1] to more general situations and to seek for an unifying idea behind his constructions. (By the way, we quote references by enclosing them in square brackets, the roman numeral indicating the chapter. Also, to refer to results within the book we first indicate the chapter, then the section and afterwards the appearance number within the section). In [IV-1] he establishes a connection between Hamiltonian flows of trivially integrable Hamiltonian systems and generalized Markov processes with independent increments.

The connection is made by developing an operational calculus quite similar to that used by physicists doing quantum mechanics. The whole set-up is also applied to the study of families of special polynomials - moment systems, naturally associated with each system (or Hamiltonian).

To extend Feinsilvers work we proceed as follows. First we develop a theory of non-unitary representations of canonical transformations. This theory suggests the "right" operational calculus needed to go from Hamiltonian flows to evolution semigroups.

Before describing the contents of the book an explanation of the title is in order. The reader may wonder how semigroups appear when representing Hamiltonian flows which have group structure. One reason for the name is the weight of tradition and the second is the fact the class of functions on which we can represent the flow by a group of integral operators is smaller than the class on which we obtain only a semigroup for positive times.

The easiest example is provided by the familiar Heath semigroup

$$P_t f(q) = \int \exp[-ikq - k^2 t/2] \hat{f}(k) \frac{dk}{(2\pi)} = \int \exp[-(q - q')^2/2t \, f(q') \frac{dq'}{(2\pi\tau)^{1/2}}.$$

Notice that the first representation is well defined for all values of t in as much as $\hat{f}(k)$ (the Fourier transform of $f(q)$) vanishes outside a bounded set (or decays faster than the exponential for large values of k), while the second representation works well for positive values of t and for any bounded function $f(q)$.

Let us briefly describe the contents of each chapter. In chapter II, the very basics of Hamiltonian mechanics are presented. We describe the class of canonical transformations that we are to deal with: those of the second type, and introduce a composition law for the generating functions which is fundamental for the representation theory to be developed later.

It is also shown that all Hamiltonian systems on $\mathbb{R}^{2n}$ for which the Hamilton-Jacobi function exists are canonically equivalent to each other and to trivially integrable systems in particular.

In chapter III a theory of non-unitary representations of canonical transformations developed. This theory, which is suggested by considerations of the simplest examples, tur out to be useful for transforming semigroups and special polynomials associated wi different Hamiltonians onto each other. At the end of the chapter we quickly draw th parallel with quantum mechanics.

In chapter IV and operational calculus, based on the representation theory introduced chapter III, is developed. These results allow us to extend Feinsilver's results to more gener situations. Here we make use of ideas introduced to describe the "correspondence rules" quantum mechanics, that is, rules describing the correspondence between functions define on phase space and operators defined on functions (or distributions) depending on th coordinates only. We also study how the correspondence rules behave under canonic transformations. We obtain as well some exponential formulae which we will employ belo quite a bit.

It is in chapter V that we take up the extension of Feinsilver's work. To eac Hamiltonian of the form $K(p) + V(q)$ we associate a semigroup with infinitesimal generat $G = K(\nabla) + V(q)$. A key ingredient is the notion of a vacuum: A positive function $\Omega($ such that $G\Omega(q) = 0$. There are some cases in which we can associate a "creation operato $C^+(t)$ to the position $q(t)$ along the trajectory of the Hamiltonian system, so that

$$f(C^+(t))1(q) = \Omega_0(q)^{-1} (e^{tG} f\, \Omega_0)(q)$$

on appropriate functions $f(q)$. This, we hope, is an improvement over the half bake treatment in [III-1].

In chapter VI we develop a variety of examples that illustrate the results and methods the previous chapters. The first five sections are computational. In the sixth we red reference [VI-3] which motivated the present line of work. In sections seven and eight w do polynomial sequences. Section 9 is devoted to taking another shot at the Heisenberg Weyl algebra and its representation as a group of canonical transformations. In section 1 we explore the connection between gauge transformations and the Cameron-Martin Girsanov formula and in the last section we play with an extended phase space under th pretext of representing the invariance group of a free Newtonian particle.

We shall now mention some commonly employed but not explicitly defined notation conventions used throughout. We begin with $\equiv$ which is taken to mean that one side of (usually the right) defines the symbol on the other (usually the left). For vectors b and $a \cdot b$ denotes their standard Euclidean scalar product. For a function $f(q)$, $\nabla f(q)$ denote the vector with components $\partial f(q)/\partial q_i$.

We also employ the multi-index notation. Multi-indices will be vectors $\mathbf{m} = (m_1,\ldots,m_n$ with integral coefficients. For a vector a in $\mathbb{R}^n$, $a^{\mathbf{m}}$ shall mean the number $a_1^{m_1} a_1^{m_2} \ldots a_1^{m}$ By $\binom{\mathbf{m}}{\mathbf{k}}$ we shall mean $\binom{m_1}{k_1} \cdots \binom{m_n}{k_n}$; $\mathbf{k}! = k_1!\, k_2! \ldots k_n!$ The inequality $\mathbf{m} \le \mathbf{k}$ mean $m_i \le k_i$ for all $i = 1,2,\ldots,n$. Also $\mathbf{1} = (1,1,\ldots,1)$.

Chapter II
The Basics of Hamiltonian Mechanics

II.1 Canonical Equations of Motion

There are many different approaches to classical (Hamiltonian) mechanics to account for different tastes and needs. A small sample list of "standard" references appears at the end of the chapter.

Classical mechanics is the mathematical apparatus constructed to describe systems of interacting particles. We shall consider the case of particles whose positions are described by points in $\mathbb{R}^n$. The Hamiltonian formalism is that presentation of classical mechanics in which there is no conceptual difference between coordinates and momenta. This becomes clear when dealing with canonical transformations, of which we shall do a lot below.

The basic starting point is the specification of a Hamiltonian function

$$H : \mathbb{R}^{2n} \times \mathbb{R} \to \mathbb{R} \tag{1.1}$$

which in simple cases corresponds to the standard mechanical energy. We shall usually write $H(q,p,t)$ or $H(q,p)$ and we adhere to the custom of denoting $q = (q_1,...,q_n)$ and $p = (p_1,...,p_n)$ as coordinates and momenta respectively. This separation of the coordinates of $\mathbb{R}^{2n}$ into two groups is via the equations of motion (1.2). When $\frac{\partial H}{\partial t} \equiv 0$ we shall say that the system is autonomous.

The motions of the system are given by the solutions to the so-called canonical or hamiltonian equations of motion:

$$\dot{q}_i = \frac{\partial H}{\partial p_i} \tag{1.2-a}$$

$$i = 1,2,...,n$$

$$\dot{p}_i = -\frac{\partial H}{\partial q_i} \tag{1.2-b}$$

which we shall assume to exist for all t, for every initial point (q,p) in $\mathbb{R}^{2n}$.

For the applications that we shall develop below, we shall consider Hamiltonians of the form

$$H(x,p) = T(p) + a(q) \cdot p + V(q) \tag{1.3}$$

or those reducible to such form by means of canonical transformations. We shall furthermore assume that $a(q) = \nabla A(q)$ for some $A : \mathbb{R}^n \to \mathbb{R}$ and we are dropping any explicit reference to the time variable in the functions T, A, V.

The standard problem in classical mechanics consists either in solving the set (1.2) exactly or approximately, or in making qualitative analyses of the solutions. However, this is not what we are going to be concerned with here.

II.2 Canonical Transformations

The possibility of making changes of variables under which the form of the canonical equations of motion is preserved is one of the most interesting aspects of the Hamiltonian formalism. (The most interesting if it were not for the curious subordination of the Hamiltonian formulation to quantum mechanics).

Once you know how to apply canonical transformations, you can search for transformations bringing (1.2) to simple form in which the analysis of the solutions is easy.

The canonical transformations that we shall consider are those termed *of the second type* in the literature, and are usually (sub-) indexed by a *2*. Since this is the only type we are going to deal with in here, for us sub-indices will always denote different transformations of the second kind.

We shall consider canonical transformations globally defined according to

Definition 2.1 The function $F(q,P,t) : \mathbb{R}^{2n+1} \to \mathbb{R}$ is said to be the *generating function* of the canonical transformation $(q,p) \to (Q,P)$ if the *transformation equations*

$$Q_i = \frac{\partial F}{\partial P_i} \tag{2.2-a}$$

$$i = 1,2,\dots,n$$

$$p_i = \frac{\partial F}{\partial q_i} \tag{2.2-b}$$

can be globally solved for (Q,P) in terms of (q,p) and (vice-versa), and if

$$\sum_i P_i \, dQ_i - \tilde{H}\, dt = \sum p_i \, dq_i - H\, dt - dF. \tag{2.3}$$

Here $\tilde{H}$ is a function of (Q,p) and perhaps t. Actually $\tilde{H}$ can be determined from

(2.3). It is given by

$$\tilde{H}(Q,P) = H(q(Q,P), p(Q,P)) + \frac{\partial F}{\partial t}(q(Q,P),P). \tag{2.4}$$

In order to deal properly with the presence of t in the equations of transformation we would have to work with an extended phase space formulation of classical mechanics. (See [1], [3], [5], [10]) for examples). To avoid overburdening the description, and since we do not allow changes in the time scale, we choose to interpret (2.2) according to physical context.

A sufficient condition for the system (2.2) to be solvable is contained in

Lemma 2.5 *A sufficient condition for* $F(q,P,t)$ *to define a canonical transformation through* (2.2) *is that*

$$\det\left(\frac{\partial^2 F}{\partial q_i \partial P_j}\right) \neq 0.$$

Comment (This condition is also necessary, but the proof is harder).

Proof Differentiate (2.2) with respect to q_j and p_i; respectively and rewrite the result in matrix form as

$$\begin{pmatrix} I & -\dfrac{\partial^2 F}{\partial P \partial P} \\ 0 & -\dfrac{\partial^2 F}{\partial P \partial q} \end{pmatrix} \begin{pmatrix} \dfrac{\partial Q}{\partial q} & \dfrac{\partial Q}{\partial p} \\ \dfrac{\partial P}{\partial q} & \dfrac{\partial P}{\partial p} \end{pmatrix} = \begin{pmatrix} \dfrac{\partial^2 F}{\partial q \partial P} & 0 \\ -\dfrac{\partial^2 F}{\partial q \partial q} & I \end{pmatrix}$$

where each of the symbols represents the obvious $n \times n$-matrix. If the hypothesis holds, then the Jacobian $\dfrac{\partial(Q,P)}{\partial(q,p)}$ is always non-vanishing and the transformation is well defined.

Since coordinate transformations have an obvious noncommutative and associative group structure this must be somehow reflected at the level of their generating functions. Let us begin with

Lemma 2.6 *Let* $F_1(q,p^{(1)})$ *and* $F_2(q^{(1)},p^{(2)})$ *be the generating functions of the canonical transformations* $(q,p) \to (q^{(1)},p^{(1)})$ *and* $(q^{(1)},p^{(1)}) \to (q^{(2)},p^{(2)})$ *respectively. Assume that*

$$q_i^{(1)} = \frac{\partial F_1}{\partial p_i^{(1)}}, \quad p_i^{(1)} = \frac{\partial F_2}{q_i^{(1)}}, \quad i = 1,2,\ldots,n \tag{2.7}$$

can be solved for $q^{(1)}, p^{(1)}$ *in terms of* $q, p^{(2)}$. *Then*

$$F(q,p^{(2)}) = F_2 \circ F_1(q,p^{(2)})$$

$$\equiv F_1(q,p^{(1)}) - q^{(1)} \cdot p^{(1)} + F_2(q^{(1)}, p^{(2)})$$

generates the composite transformation $(q,p) \rightarrow (q^{(2)},p^{(2)})$.

Proof Differentiate (2.8) with respect to $q^{(1)}$ and $p^{(1)}$ and note that (2.2) yields that $F(q,p^{(2)})$ does not depend on these variables. Differentiate with respect to q and (or) $p^{(2)}$ and obtain the correct transformation equations for F.

Comments

(i) In $F_2 \circ F_1$ the F_1 is applied first and then F_2 is applied.

(ii) The explicit reference to t is dropped.

(iii) Equations (2.7) are to be used to eliminate $q^{(1)}$ and $p^{(1)}$ in (2.8).

Let us now verify that the conditions of lemma (2.5) do in fact yield (2.7) and that we can safely go on.

Lemma 2.9 *If* $\det\left(\frac{\partial^2 F_1}{\partial q \, \partial p^{(1)}}\right) \neq 0$ *and* $\det\left(\frac{\partial^2 F}{\partial q^{(1)} \partial p^{(2)}}\right) \neq 0$ *then* $\frac{\partial(q^{(1)}, p^{(1)})}{\partial(q,p^{(2)})} \neq 0$

everywhere and the implicit function theorem can be applied to (2.7).

Proof Differentiate each of the equations of (2.7) with respect to q and $p^{(2)}$ and rewrite everything as

$$\begin{pmatrix} I & -\dfrac{\partial^2 F_1}{\partial p^{(1)}\partial p^{(1)}} \\ -\dfrac{\partial^2 F_2}{\partial q^{(1)}\partial q^{(1)}} & I \end{pmatrix}\begin{pmatrix} \dfrac{\partial q^{(1)}}{\partial q} & \dfrac{\partial q^{(1)}}{\partial p^{(2)}} \\ \dfrac{\partial p^{(1)}}{\partial q} & \dfrac{\partial p^{(1)}}{\partial p^{(2)}} \end{pmatrix}$$

$$= \begin{pmatrix} \dfrac{\partial^2 F_1}{\partial q \partial q} & 0 \\ 0 & \dfrac{\partial^2 F_2}{\partial q^{(1)}\partial p^{(2)}} \end{pmatrix}$$

The determinant of the right hand side must be that of the left hand side which by hypothesis is the product of the Jacobian determinant $\dfrac{\partial(q^{(1)},p^{(1)})}{\partial(q,p^{(2)})}$ that we want with

$$\left(1 - \det\left(\frac{\partial^2 F_1}{\partial p^{(1)}\partial p^{(1)}}\right)\det\left(\frac{\partial^2 F_1}{\partial p^{(1)}\partial q^{(1)}}\right)\right)$$

We shall now compile a list of some special classes of canonical transformations which will constitute the core of the examples we shall deal with.

Lemma 2.10 *The class of canonical transformations with generating functions of the type*

$$F(q,P,t) = P \bullet \psi(q,t) + h\,(q,t)$$

form a group. Here h *and* ψ *are assumed to be twice continuously differentiable and for each* t *the inverse* ψ^{-1} *is assumed to exists and be twice continuously differentiable as well. We*

Proof Let ψ_1, ψ_2, h_1 and h_2 be as in the statement. Then

$$F_2 \circ F_1(q,p^{(2)}) = p^{(1)} \bullet \psi_1(q) + h_1(q) - q' \bullet p'$$

$$+ p^{(2)} \bullet \psi_2(q^{(1)}) + h_2(q^{(1)})$$

$$= p^{(2)} \bullet \psi_2(\psi_1(q)) + h_2(\psi_1(q)) + h_1(q)$$

where the second step follows from $q^{(1)} = \frac{\partial F}{\partial p^{(1)}} = \psi_1(q)$. (Obviously we are dropping reference to t). Given ψ and h put

$$F^{-1}(q,P) = p \bullet \psi^{-1}(q) - h(\psi^{-1}(q)) \qquad (2.11$$

then from the previous computation it follows that

$$F^{-1} \circ F(q,P) = q \bullet P \qquad \text{and} \qquad F \circ F^{-1}(q,P) = q \bullet P.$$

It is also obvious that $F(q,P) = q \bullet P$ generates the identity transformation.

In the previous identities we use the conventional identification of initial variables as lower case and final variables as upper case. Proper symbology would be

$$F^{-1} \circ F(q,p) = q \bullet q \qquad F \circ F^{-1}(Q,P) = Q \bullet P.$$

A special subgroup, that of the "gauge transformations", is generated by

$$F(q,P,t) = q \bullet P - h(q,t)$$

which maps (q,p) into $(Q,P) = (q,p+\nabla h(q,t))$. This will be connected to the Girsanov-Martin-Cameron transformation in the probabilistic set-up. This shall be explored in chapters III and IV.

The other special subgroup of canonical transformations is described in

Lemma 2.13 *The class of canonical transformations generated by*

$$G(q,P,t) = q \bullet \varphi(P,t) + h(P,t)$$

is a group. Again we assume the existence of derivatives as needed and the invertibility of φ *for each* t..

Proof We only remark that the inverse of $F(q,P) = q \bullet \varphi(P) + h(P)$ is given by

$$F^{-1}(q,P) = q \bullet \varphi^{-1}(P) - h(\varphi^{-1}(P)).$$

Lemma 2.14 *The generating function*

$$F(q,P) = \varphi(P) \bullet \psi(q) = h(q) + g(P)$$

s the result of composing $F_1(q,P) = q \cdot P + h(q)$ *with* $F_2(q,P) = q \cdot \varphi(p) + g(P)$.

Proof Homework for the reader.

1.3 The Hamilton-Jacobi Equation

To the system (1.2) one associates a flow $\varphi_t : \mathbb{R}^{2n+1} \to \mathbb{R}^{2n+1}$ given by

$$\varphi_t(p,q,s) = (T_{s,t}(p,p),t+s)$$

where $T_{s,t}(p,q)$ is the solution to (1.2) which at time s is at (q,p). Whenever H is independent of t, instead of (3.1) we can consider the flow $T_t : \mathbb{R}^{2n} \to \mathbb{R}^{2n}$ given by

$$T_t(q,p) = (q(t;q,p),\, p(t;q,p)) \equiv (\bar{q},\bar{p}) \tag{3.2}$$

where by $(\bar{q},\bar{p})$ we shall denote the coordinates at t of the trajectory which at $t=0$ is at (q,p).

It is a standard fact that $T_t : \mathbb{R}^{2n} \to \mathbb{R}^{2n}$ is a group of canonical transformations taking initial data onto present [time] data. If we think of initial data as constant trajectories of a system with constant (=0) Hamiltonian, we can ask whether there exists a generating function from which the canonical flow T_t can be obtained.

An answer provided by the Hamilton-Jacobi theory goes as follows. Let $S(\bar{q},p,t)$ denote the generating function taking the system present time t (and Hamiltonian $H(\bar{q},\bar{p})$) onto the system at rest at $t = 0$. From (2.2) we obtain that

$$\bar{p}_i = \frac{\partial S}{\partial \bar{q}_i}$$

which can be substituted into (2.4) to obtain

$$\frac{\partial S}{\partial t} + H(\bar{q}, \nabla_{\bar{q}} S) = 0 \tag{3.3}$$

which can now be thought of as an equation defining S. This equation is called the Hamilton-Jacobi equation and S is called the Hamilton-Jacobi function.

Since at $t = 0$ we want S to generate the identity transformation, we require

$$S(\bar{q},p,0) = \bar{q} \cdot p \tag{3.4}$$

to hold at $t = 0$. The same considerations as above yield that the generating function of the transformation $\bar{S}$ that takes the system from rest at $t = 0$ to the system at t will have to satisfy

$$\frac{\partial \bar{S}}{\partial t} = H(q, \nabla_q \bar{S}) \tag{3.5}$$

with the initial condition

$$\bar{S}(q, \bar{p}, 0) = q \cdot \bar{p} . \tag{3.6}$$

Comment In the pairs (3.3) - (3.4) and (3.5) - (3.6) the p's appearing in the initial condition play the role of integration constants. With this in mind we have

Lemma 3.7 *The solution to* (3.3) - (3.4) (*or* (3.5) - (3.6)) *of the type*

$$S(\bar{q}, p, t) = K(q, p) - E\,t$$

exists only when $H(q,p) = T(p)$ . .

Proof Since $S(\bar{q},p,t) = \bar{q} \cdot p$ we have $K(\bar{q},p) = \bar{q} \cdot p$ then $E = H(\bar{q}, \nabla_{\bar{q}} S) = H(\bar{q},p)$ and since the left hand side does not depend on $\bar{q}$, we are forced to have $H(\bar{q},p) = T(p)$ and then

$$S(\bar{q}, p, t) = \bar{q} \cdot p - t\,H(p)$$

solves (3.3) - (3.4).

Analogously we have

$$\bar{S}(q, \bar{p}, t) = q \cdot \bar{p} + tH(\bar{p}).$$

Definition 3.8 We shall say that a system on IR is *(trivially) integrable* if its motion is described by a Hamiltonian $H = H(p)$.

This is obviously so because in this case, (1.2) becomes $\dot{q} = \nabla_p T(p)$ and $\dot{p} = 0$, which can be trivially integrated.

We shall leave for the reader to verify that the S and $\bar{S}$ defined in lemma (3.7) are inverses to each other.

Lemma 3.9 *Assume from now on that the solution to* (3.3) - (3.4) *exists, is three times continuously differentiable and put* $\frac{\partial S}{\partial q_i} = \xi_i = \xi_i(\bar{q},p,t)$. *Then* $\xi_i = \bar{p}_i = p_i(t;q,p)$.

Proof Note that $\xi_i(\bar{q},p,0) = p_i$ and differentiating (3.3) with respect to $\bar{q}_i$, making use of (1.2), we have

$$\frac{\partial}{\partial t}\xi_i + \frac{\partial H}{\partial \bar{q}_i} + \frac{\partial H}{\partial \bar{p}_j}\frac{\partial^2 S}{\partial \bar{q}_j \partial q_i} = \frac{\partial \xi_i}{\partial t} + \dot{\bar{q}}\frac{\partial \xi_i}{\partial \bar{q}_j} - \dot{\bar{p}}$$

$$= \frac{d}{dt}(\xi_i - \bar{p}_i) = 0.$$

Since at $t = 0$ $\xi_i(\bar{q},p,0) - \bar{p}_i(0) = p_i - p_i = 0$, then $\dot{\xi}_i = \bar{p}_i$ for all t (in the interval in which S is defined, of course).

Lemma 3.10 *Under the same hypothesis of lemma 3.9, put* $\zeta_i = \dfrac{\partial S}{\partial p_i} = \zeta_i(\bar{q},p,t)$. Then $\zeta_i = p_i$.

Proof Same procedure as above. Do it!.

Lemma 3.11 *Let* $M_{ij} = \dfrac{\partial^2 S}{\partial \bar{q}_i \partial p_j}$. *Then*

$$\det M(t) = \det M(0) \exp\left\{ - \int_0^t \operatorname{tr} \tilde{R}_{ij}(s)\, ds \right\}$$

with $\tilde{R}_{ij}(s) = \dfrac{\partial^2 H}{\partial \bar{q}_i \partial \bar{p}_j}(\bar{q}(s),\bar{p}(s),s)$.

Proof Differentiate (3.3) with respect to $\bar{q}_i$ and p_j to obtain

$$\frac{\partial}{\partial t}\frac{\partial^2 S}{\partial \bar{q}_i \partial p_j} + \frac{\partial^2 H}{\partial \bar{q}_i \partial \bar{p}_k}\frac{\partial^2 S}{\partial \bar{q}_k \partial p_j} + \frac{\partial H}{\partial \bar{p}_k}\frac{\partial}{\partial \bar{q}_k}\frac{\partial^2 S}{\partial \bar{q}_i \partial p_j} = 0.$$

Put $\tilde{M}(t) = M(\bar{q}(t),t)$. From lemma (3.9) we can replace $\nabla_{\bar{q}} S$ with $\bar{p}(t)$ and obtain

$$\frac{d(\det \tilde{M})}{dt} + \operatorname{tr} \tilde{R}\, (\det \tilde{M}) = 0$$

from which the desired conclusion follows.

Comment From these three lemmas we see that whenever the solution to (3.3) - (3.4) is defined and three times continuously differentiable, then the transformation equations $\bar{p} = \nabla_{\bar{q}} S$, $q = \nabla_p S$ can be solved to obtain (q,p) in terms of $(\bar{p},\bar{q})$ since $\det\left(\frac{\partial^2 S}{\partial \bar{q}_i \partial p_j}\right) \neq 0$.

To finish this digression about S we obtain a standard representation of S. Let $(q(t),p(t))$ be the solution to (1.2) passing through (q,p) at $t = 0$. Assume that (3.3) - (3.4) has a solution in $[0,T]$, and for each $t \in [0,T]$

$$(\hat{q}(s), \hat{p}(s)) = (q(t-s), p(t-s)) \quad \text{and} \quad \hat{S}(t) = S(\hat{q}(s), p, t-s).$$

Then

$$\frac{d\hat{S}}{dt} = -\dot{\bar{q}}(t-s) \cdot \nabla_{\bar{q}} S + H(q(t-s), p(t-s)).$$

Now integrate from 0 to t, we obtain

$$S(\hat{q}(t), p, 0) - S(\hat{q}(0), p, t) = -\int_0^t (p \cdot \dot{q} - H)\, ds$$

and since $\hat{q}(t) = q$ and $\hat{q}(0) = \bar{q}$, therefore

$$S(\bar{q}, p, t) = q \cdot p + \int_0^t \{\, \bar{p}(s) \cdot \dot{\bar{q}}(s) - H(\bar{q}(s), \bar{p}(s))\,\}\, ds. \tag{3.12}$$

Recall that $q_i = \dfrac{\partial S}{\partial p_i} = q_i\,(\bar{q},p,t)$, but solving for $(q,\bar{p})$ from $\bar{q} = \bar{q}(t,q,p)$ and $\bar{p} = \bar{p}(t,q,p)$ yields $S(\bar{q},p,t)$ without having to solve (3.3).

In what remains of this section we shall consider the problem of reconstructing $S(\bar{q},p,t)$ from its infinitesimal version. To be more precise, note that from the equations of motion (1.2) it follows that up to terms of $O(\varepsilon^2)$

$$\begin{aligned} q(\varepsilon) &= q + \varepsilon \nabla_p H(q,p) \\ p(\varepsilon) &= p - \varepsilon \nabla_q H(q,p). \end{aligned} \tag{3.13}$$

Note now that the generating function

$$S(\bar{q},p) = \bar{q} \cdot p - \varepsilon H(\bar{q},p) \tag{3.14}$$

generates the transformation

$$\begin{aligned} q &= \bar{q} - \varepsilon \nabla_p H(\bar{q},p) & \qquad \bar{q} &= q + \varepsilon \nabla_p H(\bar{q},p) \\ & & \text{or} & \\ \bar{p} &= q - \varepsilon \nabla_{\bar{q}} H(\bar{q},p) & \qquad \bar{p} &= p - \varepsilon \nabla_{\bar{q}} H(\bar{q},p) \end{aligned} \tag{3.15}$$

which up to $O(\varepsilon^2)$ coincides with (3.13). In other words, the infinitesimal time evolution can be compensated by an infinitesimal canonical transformation.

Let now t be an arbitrary fixed positive real number and, again, let $(\bar{q},\bar{p})$ denote the present position of a trajectory of (1.2) which at $t = 0$ passes through (q,p). For every $n \geq 1$ put

$$q^k = q\left(t\left(1 - \frac{k}{n}\right)\right), \quad p^k = p\left(t\left(1 - \frac{k}{n}\right)\right) \quad k = 1,2,\ldots,n.$$

The (q^k,p^k) are points along the trajectory, but sampled in reverse at time intervals of length $1/n$. Let us introduce

$$S_n^{(k)}(q^{k-1},p^k) = q^{k-1} \cdot p^k - \frac{t}{n} H(q^{k-1},p^k) \tag{3.16}$$

and from the composition

$$\begin{aligned} S^{(n)}(q^0,p^n,t) &= S_n^n \circ S_n^{n-1} \circ \ldots \circ S_n^1 (q^0,p^n,t) \\ &= S_n^{(1)}(q^0,p') - q' \cdot p' + S_n^2(q',p^2) - q^2 \cdot p^2 \\ &\quad + \ldots - q^{n-1} \cdot p^{n-1} + S_n^n(q^{n-1},q^n) \end{aligned} \tag{3.17}$$

which can be rewritten as

$$S^{(n)}(q^0,p^n,t) = \sum_{k=1}^{n} p^k \cdot (q^{k-1} - q^k) - \sum_{k=1}^{n} H(q^{k-1},p^k) \frac{t}{n} + q\left(\frac{t}{n}\right) \cdot p(0)$$

and taking the limit as $n \to \infty$ we obtain that

$$S^{(n)}(q^0,p^n,t) \to q(0) \cdot p(0) + \int_0^t (p \cdot dq - H\,dt) \tag{3.18}$$

where the integral is taken along the solution to (1.2) that we started with. Solving for q in terms of $(\bar{p},p)$ we obtain

Proposition 3.19 *The Hamilton-Jacobi function can be obtained as limit of compositions of its infinitesimal versions* $S_\varepsilon(q,p)$ *given by* (3.14) *as described in* (3.17).

II.4 Canonical Equivalence of Integrable Systems on $\mathbb{R}^{2n}$

Above we defined trivially integrable systems to be those with Hamiltonian $H = H(p)$ depending on p only and, we saw in lemma (3.7) and following comments that

$$S(\bar{q},p,t) = \bar{q} \cdot p - t\,H(p)$$

$$\bar{S}(q,\bar{p},t) = q \cdot \bar{p} + t\,H(p)$$

are inverses to each other under the composition law (2.8).

Proposition 4.1 *Consider two integrable systems on* $\mathbb{R}^{2n}$ *with Hamiltonians* $H_1(p)$ *and* $H_2(P)$ *respectively. Then the canonical transformation generated by*

$$F(q,P,t) = q \cdot P - t\,H_1(P) + t\,H_2(P)$$

maps the system with Hamiltonian $H_1(p)$ *onto the system with Hamiltonian* $H_2(P)$.

Proof It is trivial, but let us stretch it out to see what lies behind the result. Notice that the $F(q,P,t)$ given in (4.2) is the composition of $S_1(\bar{q},p,t) = \bar{q} \cdot p - t\,H_1(p)$ and $\bar{S}_2(q,P,t) = q \cdot P + t\,H_2(P)$. The first one brings the system evolving according to H_1 back to rest and the second brings the system from rest to present time according to H_2.

A slight extension is contained in the following result which asserts that when $S(\bar{q},p,t)$ and $\bar{S}(q,\bar{q},t)$ exist, the system is integrable.

Proposition 4.3 *Let us assume that* $H(q,p)$ *is such that* $S(\bar{q},p,t)$ *and* $\bar{S}(q,\bar{p},t)$ *given by* (3.3) - (3.5) *exist. Then this system is equivalent to any trivially integrable system on* $\mathbb{R}^{2n}$ *with Hamiltonian* $\tilde{H}(p)$, *and the equivalence is brought about by*

$$F(\bar{q},P,t) = \tilde{S} \circ S(\bar{q},P,t) \tag{4.4}$$

with $\tilde{S}(q,P,t) = q \cdot P + t\,\tilde{H}(P)$.

Proof Just verify that $F(\bar{q},P,t) = S(\bar{q},p,t) - q \cdot p + \tilde{S}(q,P,t)$ does the job. We can simplify this identity by noting that $p = P = \nabla_q \tilde{S}$ and obtain

$$F(\bar{q},P,t) = \tilde{S}(\bar{q},P,t) + \tilde{H}(P)\,t.$$

We leave for the reader to verify that

$$\frac{\partial F}{\partial t} + H(\bar{q},P) = \tilde{H}(P).$$

As an illustration consider the transformation

$$F(q,P,t) = \frac{q\,P}{\cos t} - \frac{q^2 + P^2}{2}\tan t + \frac{t\,P^2}{2}$$

mapping the one dimensional oscillator with Hamiltonian $\frac{(p^2+q^2)}{2}$ onto a system (the free particle) with Hamiltonian $\frac{p^2}{2}$.

Notice that even though F has a singularity at $t = 2\pi$, the equations of transformation it defines present no problem. The correct set up to deal with these transformations is the extended phase space in which the time variable is added as an extra coordinate. In this set up no mysterious behaviour of phase-space trajectories appears.

Also, the results in [7] asserting that every quadratic Hamiltonian can be written as a sum of Hamiltonians of type $\frac{(p^2 + w^2 q^2)}{2}$, taken jointly with the previous example shows how to trivialize systems with quadratic Hamiltonians.

II.5 Two Time Hamilton-Jacobi Functions and Time Evolution

The basic property of the Hamilton-Jacobi function introduces a time asymetry into our scheme. Note that for autonomous systems, the flow $T_t : \mathbb{R}^{2n} \to \mathbb{R}^{2n}$ determines a one parameter family of canonical transformations $(q,p) \to (\bar{q},\bar{p}) = (q(t;q,p),p(t;q,p))$ such that $H(q,p) = H(\bar{q},\bar{p})$.

It is clear that the time at which the initial coordinates is specified is arbitrary. Denote it by t_0.

If we denote by $S(t+t_0,\bar{q},P)$ the generating function bringing the system from time $t+t_0$ to rest at 0 and by $\bar{S}(t_0,Q,p)$ the transformation taking the system from rest to present state at t_0, then

$$W(t+t_0,\bar{q};t_0,p) = S(t+t_0) \circ \bar{S}(t_0)(\bar{q},p) \tag{5.1}$$

denotes the transformation taking the system from $t+t_0$ to t_0.

Observe that if the running variable is taken to be t then

$$\frac{\partial W}{\partial t_0}(t+t_0,\bar{q};t_0,p) = \frac{\partial S}{\partial t_0}(t+t_0;p,P) - Q \cdot P + \frac{\partial \bar{S}}{\partial t_0}(Q,p) = -H(\bar{q},\bar{p}) + H(q,p)$$

where we made use of $\frac{\partial S}{\partial t}(t+t_0;q,P) = -H(\bar{q}, \nabla_{\bar{q}}S) = -H(\bar{q},\bar{p})$ and $\frac{\partial \bar{S}}{\partial t} = H(q,\nabla_q\bar{S}) = H(q,p)$, and therefore

$$\frac{\partial W}{\partial t_0}(\ ) + H(\bar{q},\bar{p}) = H(q,p) \tag{5.2}$$

where the $(\bar{q},\bar{p})$ in the left hand side have to be written in terms of (q,p) using the transformation equations

$$\frac{\partial W}{\partial \bar{q}} = \bar{p} \qquad \frac{\partial W}{\partial p} = q.$$

Let us now go back to the problem of the choice of the initial starting time t_0. When the Hamiltonian $H(q,p)$ is time independent the system is autonomous, which means that given $(q(t+t_0),p(t_0))$ the state $(q(t+t_0,p(t_0+t))$ depends only on t. This means that $W(t+t_0,\bar{q};t_0,p)$ and the composition on the right hand side of (5.1) does not depend on t_0. This is correctly taken care of in (5.2) since $H(\bar{q},\bar{p}) = H(q,p)$.

The proper discussion of this issue involves a formulation of Hamiltonian theory in an extended phase space which includes time as an extra coordinate.

The following example will appear naturally in the next section. Let $H(q,p) = H(p)$ only. Then

$$S(\bar{q},P,t) = \bar{q} \cdot P - t\,H(P) \quad \text{and} \quad \tilde{S}(t,Q,p) = Q \cdot p + t\,H(p)$$

and therefore

$$W(t+t_0,\bar{q},t_0,p) = \bar{q}\cdot P - (t+t_0)\,H(\bar{P}) - Q\cdot P +$$

$$+ Q\cdot p + t_0\,H(P) = \bar{q}\cdot p - tH(p)$$

ince in this case $\bar{p} = P = p$. We see that $W(t+t_0,\bar{q};t_0,p)$ does not depend explicitly on t_0 but it is needed to get (4.2)) and furthermore that $W(t+t_0,\bar{q};t_0,p) = S(t,\bar{q},p)$!

1.6 Poisson Brackets and Infinitesimal Canonical Transformations

For any two functions $f(q,p)$ and $g(q,p)$ defined on $\mathbb{R}^{2n}$ we define their Poisson bracket $[f,g]$ by

$$[f,g] = \sum_i \frac{\partial f}{\partial q_i}\frac{\partial g}{\partial p_i} - \frac{\partial f}{\partial p_i}\frac{\partial g}{\partial q_i}\,. \tag{6.1}$$

With this definition, the equations of motion (1.2) can be rewritten as

$$\dot{q}_i = [q_i,H], \quad \dot{p}_i = [p_i,H], \quad i = 1,2,\ldots,n \tag{6.2}$$

and the change in time of any function $f(q,p,t)$ along the solution to (6.2) is given by $\frac{df}{dt} = \frac{\partial f}{\partial t} + [f,H]$. Verify it!.

The notion of Poisson bracket is especially handy for defining the canonically of a change of variable.

We shall say, and this is the general definition, that a transformation $(Q,P) = (Q(q,p,t), P(q,q,t))$ is canonical whenever it leaves the Poisson brackets invariant, i.e.

$$[Q_i,P_j]_{Q,P} \equiv \sum \frac{\partial Q_i}{\partial Q_k}\frac{\partial P_j}{\partial P_k} - \frac{\partial Q_i}{\partial P_k}\frac{\partial P_j}{\partial Q_k}$$

$$= \sum_k \frac{\partial Q_i}{\partial q_k}\frac{\partial P_j}{\partial p_k} - \frac{\partial Q_i}{\partial p_k}\frac{\partial P_j}{\partial q_k} \equiv [Q_i,P_j]_{q,p} = \delta_{ij} \tag{6.3-a}$$

and analogously

$$[Q_i,Q_j]_{Q,P} = [Q_i,Q_j]_{q,p} = [P_i,P_j]_{Q,P} = [P_i,P_j]_{q,p} = 0$$

where the sub-indices to the bracket symbol indicate with respect to which variables the derivations are carried out.

A family $(Q_s,P_s) = (Q(q,p,s),P(q,p,s))$ of canonical transformations is called a one parameter group of canonical transformations whenever

$$(Q_{t+s},P_{t+s}) = (Q(Q_s,P_s,t),Q(Q_s,P_s,t))$$

for any s, t and, for consistency $(Q_0,P_0) = (q,p)$.

We shall see how to obtain a one parameter group of canonical transformations from i infinitesimal version. For s small we have $Q_i = q_i + s\, f_i\,(q,p)$, $P_i = p_i + s\, f_{n+i}(q,p)$ up t order s^2 where $f_i(q,p) = \frac{dQ_i}{ds}|_{s=0}$, $f_{n+i}(q,p) = \frac{dP_i}{ds}|_{s=0}$. Differentiating (6.3-a) and (6.3-b with respect to s and computing at s=0 we obtain as result that there exists a functio $f(q,p)$ such that at s=0

$$\frac{dQ_i}{ds} = [\,Q_i,f\,] \quad \text{and} \quad \frac{dP_i}{ds} = [\,P_i,f\,]$$

The set (6.4) is the obvious generalization of (1.2), or to put it otherwise, tim evolution is just one more case of a one parameter group of canonical transformations.

To come back to our case, i.e. that of canonical transformations with generating functio $F(q,P,s)$, when s is near zero, $F(q,P,s)$ must be near the generating function of the identit transformation and we have

$$F(q,P,s) = q \bullet p + s\, f(q,P) + O(s^2) \qquad (6.5$$

with

$$f(q,P) = \frac{\partial F(q,P,s)}{\partial s}|_{s=0}\,. \qquad (6.6$$

We leave it for the reader to verify that (6.4) is obtained when we apply th transformation equations (2.2) to (6.5).

An analogous reasoning to that carried out for the Hamilton–Jacobi equation in section 3 shows that the solution to

$$\frac{\partial \Sigma(q,P,s)}{\partial s} + f(q,\nabla_q \Sigma) = 0, \qquad \Sigma(q,P,0) = q \cdot P \qquad (6.7$$

is the generating function for the canonical transformation mapping $(Q(s),P(s))$ back ont (q,p). This $\Sigma(q,P,s)$ will play an important role in chapter IV.

;eferences

ooks on Classical mechanics

[I-1] Thirring, W. : "Classical Dynamical Systems". Springer-Verlag, Berlin, 1981.

[I-2] Gallavotti, G. : "The Elements of Mechanics". Springer-Verlag, Berlin, 1983.

[I-3] Whittaker, E. T. : "A Treatise on the Analytical Dynamics of Particles and Rigid Bodies". Cambridge University Press, Cambridge, 1964.

[I-4] Arnold, V. : "Méthodes Mathematiques de la Mechanique Classique". Editions MIR, Moscow, 1976.

description of canonical transformations in extended phase space can be seen in

[I-5] Asorey, M., Cariñena, J. F. and Ibort, L. A. : "Generalized Canonical Transformations for Time Dependend Systems". Jour. Math. Phys. **24**, No. 2 (1982) 2745 - 2760.

[I-6] Kuwabara, Ruishi : "Time Dependent Mechanical Symetries and Extended Hamiltonian Systems". Reps. Math. Phys. **19**, No. 1 (1984) 27 - 38.

reduction of quadratic Hamiltonians to a common case appeared in

[I-7] Bogdanovich, R. and Gopinathan, M. S. : "A Canonical Transformation of the Hamiltonians Quadratic in Coordinate and Momentum Spaces". J. Phys. A. Math. Gen. **12**, No. 9, (1979) 1457 - 1467.

classification of quadratic hamiltonians is contained in

[I-8] Williamson, J. : "On the Algebraic Problem Concerning the Normal Forms of Linear Dynamical Systems". Amer. Jour. Mathem. **58** (1936) 141 - 163.

Chapter III
Representation of Canonical Transformations

III.1 The Representation Formula

In this section we shall develop a theory of nonunitary representations of the two subgrou of canonical transformations introduced in section II. Using this theory we are going to carr out a program analogous to what is done in quantum mechanics, i.e., we will associa evolution (semi) groups to Hamiltonian flows and construct an operational calculus.

To motivate our definitions consider the coordinate transformations $q \to Q = \varphi($ generated by $F(q,P) = P \cdot \varphi(q)$. If $\tilde{f}(Q)$ is a given function, its transform should be $f(q)$ $\tilde{f}(\varphi(q))$.

We want this to be the result of some integral operator T_F^*, but we want it to be define in a class of functions as large as possible. Thus we will obtain $T_F^* f$ by an appropria dualization on operators well defined on test functions.

The integral operators we will define throughout are to act on functions F defined on such that

$$\hat{\Phi}(k) = \int \exp(ik \cdot \xi)\, \Phi(\xi)\, d\xi$$

has compact support. From this we can extend to functions such that $\hat{\Phi}(k)$ is of slo growth.

Test functions Φ can be recovered from their Fourier transforms by means of

$$\Phi(\xi) = \int \hat{\Phi}(\xi) \exp(-ik \cdot \xi) \frac{dk}{(2\pi)^n}.$$

Here we follow the opposite the usual conventions regarding Fourier transforms that are use in quantum mechanics and probability. All over the place we shall make use of the integr representation of the -"δ-function"

$$\delta(q - q') = \int \exp(i(q - q')k) \frac{dk}{(2\pi)^n}. \quad (1.$$

Let now $\hat{\Phi}(Q)$ be a test function of the Q-variables and let $F(q,P) = P \cdot \varphi(Q)$, and let $f($

: a given function, and let $\tilde{f}(Q) = f(\varphi^{-1}(Q)) = (T_F^* f)(Q)$ be the obvious composition. We ant to define $T_F\tilde{\Phi}$ in such a way that

$$\int \tilde{f}(Q)\,\tilde{\Phi}(Q)\,dQ \equiv \langle T_F^* f, \tilde{\Phi} \rangle = \langle f, T_F\tilde{\Phi} \rangle = \int f(q)\,\Phi(\varphi(q))\,J(q)\,dq$$

› the choice for $(T_F\tilde{\Phi})(q)$ is $(T_F\Phi)(q) = \tilde{\Phi}(\varphi(q))\,J(q)$ with $J(q)$ being the determinant of ıe matrix $\frac{\partial\varphi_i(q)}{\partial q_j}$.

Note now that $J(q)$ is also the determinant of $\frac{\partial^2 F}{\partial q_j \partial P_i}$ and that $\tilde{\Phi}(\varphi(q))$ can be written as

$$\int \exp(-ik \,.\, \varphi(q))\,\tilde{\Phi}(k)\,\frac{dk}{(2\pi)^n}.$$

/e proceed to state

efinition 1.2 Let the generating function $F(q,P)$ be such that $F(q,P)$ is well defined, then 'e put for $\tilde{\Phi}(Q)$ above

$$(T_F\tilde{\Phi}(q) = \int \Delta(q,ik)\exp(-F(q,ik))\,\hat{\Phi}(k)\,\frac{dk}{(2\pi)^n} \qquad (1.3)$$

'omments The symbol $F(q,ik)$ will make sense in a variety of circumstances, for example, hen $F(q,P)$ is analytic in P. We are also not worrying about the dependence of F in t for ıe time being.

emma 1.4 *Let* $F_i(q,P) = P \bullet \varphi_i(q) + h_i(q)$ *be generating functions of canonical ansformations. Then*

$$T_{F_2 \circ F_1}\tilde{\Phi} = T_{F_1} T_{F_2}\tilde{\Phi}. \qquad (1.5)$$

roof Begin by applying (1.3) to note that for $F(q,P) = P \bullet \varphi(q) + h(q)$ we obtain

$$(T_F\tilde{\Phi})(q) = \exp(-h(q))\,\tilde{\Phi}(\varphi(q))\,J(q)$$

ıen

$$(T_{F_1} T_{F_2} \tilde{\Phi})(q)$$

$$= \exp(-h_1(q)) \exp(-h_2(\varphi_1(q))) \tilde{\Phi}(\varphi_2(\varphi_1(q))) J_2(\varphi_1(q)) J_1(q).$$

Note that the Jacobian $J(q)$ of $\varphi_2 \circ \varphi_1(q)$ is $J_2(\varphi_1(q)) J_1(q)$ and that $F_2 \circ F_1(q,P) =$ $P \bullet \varphi_2(\varphi_1(q)) + h_1(q) + h_2(\varphi_1(q))$.

Comments It is clear that the correspondence $F \to T_F$ is a non-unitary anti-representation. This qualifier refers only to the order in which one applies the operators T_F relative to the order in which the functions φ are applied.

To represent $F(q,P) = q \cdot \varphi(P)$, with P a smooth invertible mapping on $\mathbb{R}^n$, we have two alternatives: either by imposing extra conditions on $\varphi(P)$ or by stretching representation (1.1).

Proposition 1.6 *Let* $F_l(q,P) = q \cdot \varphi_l(P)$, $l = 1,2$ *be diffeomorphisms such that there exist a function* $\tilde{\varphi}_1$ *such that*

$$i\,\varphi_l(k) = \varphi_l(ik) \qquad l = 1,2.$$

Then (1.5) *holds as well when* $T_F \tilde{\Phi}$ *is defined by* (1.3).

Proof In this case $\Delta = \det \left| \dfrac{\partial^2 F}{\partial P_i \partial q_j} \right| = \det \left| \dfrac{\partial \varphi_j}{\partial P_j} \right|$ and

$$(T_F \tilde{\Phi})(q) = \int \Delta(ik) \exp(-q \cdot \varphi(ik)) \hat{\Phi}(k) \frac{dk}{(2\pi)^n}$$

$$= \int \exp(-iq \cdot k) \hat{\hat{\Phi}} \left(\frac{\varphi(ik)}{i} \right) \frac{dk}{(2\pi)^n}$$

where we made the change of variable $ik = \varphi^{-1}(i\bar{k})$ and then renamed $\bar{k}$ to be k again. The upshot of the computation is that

$$(T_F \tilde{\Phi})^\wedge (k) = \hat{\hat{\Phi}} \left(\frac{\varphi(ik)}{i} \right) = \hat{\hat{\Phi}} (\tilde{\varphi}^{-1}(k)).$$

From this it follows that

$$(T_{F_1} T_{F_2} \tilde{\Phi})^\wedge (k) = \hat{\tilde{\Phi}}(\tilde{\varphi}_2^{-1} (\varphi_1^{-1}(k)))$$

nd taking into account that $F_2 \circ F_1(q,P) = q \cdot \varphi_1(\varphi_2(P))$ we obtain our desired result.

When $\frac{\varphi(ik)}{i}$ is not real but $\varphi(ik)$ is well defined then so is $\tilde{\Phi}\left(\frac{\varphi(ik)}{i}\right)$. Actually this amounts to saying that

$$\int \exp (iqk + q \cdot \overline{\varphi(ik)}) \frac{dq}{(2\pi)^n} = \int \exp (iq \cdot (k - \varphi(ik)i) \frac{dq}{(2\pi)^n}$$

$$\equiv \delta(k - i\varphi(ik))$$

which we shall use now and then.

Let us now examine the effect of $F(q,P) = q \cdot P + h(P)$ for analytic $h(P)$.

Proposition 1.7 *Let* $F_i(q,P) = q \cdot P + h_i(P)$. *Then* (1.5) *is again valid.*

Proof In this case $\Delta = 1$ and

$$(T_F \tilde{\Phi})(q) = \int \exp (- F(q,ik)) \hat{\Phi}(k) \frac{dk}{(2\pi)^n} = \exp (- h(D) \tilde{\Phi}(q)$$

Therefore, since $F_2 \circ F_1(q,P) = q \cdot P + h_1(P) + h_2(P)$ the result is clear.

Comment Since $F(q,P) = q \cdot \varphi(P) + h(P)$ is the composition, $F_1(q,P) = q \cdot \varphi(P)$ and $F_2(q,P) = q \cdot P + h(P)$, $T_F = T_{F_1} T_{F_2}$ gives the action of T_F.

Another way to look at the effect of T_F for $F(q,P) = q \cdot \varphi(P)$ is the following

$$(T_F \tilde{\Phi}(q) = \int \exp (- q \cdot \varphi(ik)) \Delta(ik) \hat{\tilde{\Phi}}(k) \frac{dk}{(2\pi)^n}$$

$$= \int \Delta(ik) \exp (-q \cdot \varphi(ik)) \left(\int \exp (i\tilde{k} \cdot Q)\tilde{\Phi}(Q)\, dQ\right) \frac{dk}{(2\pi)^n}$$

$$= \int \Delta(ik) \left(\int (\exp(-q \cdot \varphi(D))) \exp(ik \cdot Q)) \tilde{\Phi}(Q)\, dQ \right) \frac{dk}{(2\pi)^n}$$

$$= \int \Delta(ik) \left(\int \exp(ik \cdot Q) (\exp(-q \cdot \varphi(D)) \tilde{\Phi})(Q)\, dQ \right) \frac{dk}{(2\pi)^n}$$

$$= \int \left(\int \exp(ik \cdot Q) (\Delta(-D) \exp(-q \cdot \varphi(-D)) \tilde{\Phi})(Q)\, dQ \right) \frac{dk}{(2\pi)^n}$$

$$= \int \frac{dk (\Delta(-D) \exp(-q \cdot \varphi(-D)) \tilde{\Phi})^{\wedge}(k)}{(2\pi)^n}$$

$$= (\Delta(-D) \exp(-q \cdot \varphi(-D)) \bar{\Phi})(0).$$

In the proofs of propositions (1.6) and (1.5) we obtained the following relations for th Fourier transforms of $(T_{F_1} \tilde{\Phi})$, , $l = 1,2$, with $F_1 = q \cdot \varphi(P)$ and $F_2 = q \cdot P + h(P$ respectively we have

$$(T_{F_1} \tilde{\Phi})(k) = \hat{\tilde{\Phi}}\left(\frac{\varphi(ik)}{i}\right)$$

(1.

$$(T_{F_2} \tilde{\Phi})(k) = \exp h(ik)\, \hat{\tilde{\Phi}}(k).$$

Lemma 1.9 *Let* $F(q,P) = \varphi(q) \cdot \psi(P) + h(P) + g(q)$ *be the generating function of canonical transformation, then* T_F *can be computed as* $T_{F_2 \circ F_1}$ *with* $F_1(q,P) = P \cdot \varphi(q)$ $g(q)$ *and* $F_2(q,P) = q \cdot \psi(P) + h(P)$. *Here* φ, ψ, h, g *are to satisfy all the requirements w stated in lemmas* (1.5) *and* (1.6).

Proof We leave it for the reader to work out and to verify whether

$$(T_F \tilde{\Phi})(q) = \int \Delta(q,k) \exp -F(q,ik)\, \hat{\tilde{\Phi}}(k) \frac{dk}{(2\pi)^n}$$

holds true.

To close this section, we mention that unless $F(q,P) = q \cdot P$ it would happen that

$$(T_F \tilde{\Phi}_1, T_F \tilde{\Phi}_2) \neq (\tilde{\Phi}_1, \tilde{\Phi}_2) = \int \tilde{\Phi}_1(Q)\, \tilde{\Phi}_2(Q)\, dQ$$

hich is the meaning of non-unitarity of the integral transforms we have defined so far.

.2 Canonical Transformations of Distributions

et us briefly examine some regularity properties of the operators T_F corresponding to the sic generating functions introduced in chapter II. We shall recall that a C^∞ function $\tilde{\Phi}$ fined on $\mathbb{R}^n$ is of slow growth (or *in S*) whenever for any integer p and any multi- dex α of length $|\alpha| \leq p$ we have $|Q|^p |D^\alpha \tilde{\Phi}(Q)| \to 0$ as $|Q| \to \infty$, or equivalently, if r any integer p

$$\sup_{\substack{Q \in \mathbb{R}^n \\ |\alpha| \leq p}} (1 + |Q|^2)^{p/2} |D^\alpha \tilde{\Phi}(Q)| < \infty. \tag{2.1-a}$$

Recall as well that a function $g(q)$ in $C^\infty(\mathbb{R}^n)$ is of polynomial growth whenever for y multi-index α there exist a constant c_α and an integer m_α such that

$$|D^\alpha g(Q)| \leq c_\alpha (1 + |q|)^{m_\alpha}. \tag{2.1-b}$$

/e can proceed to

emma 2.2 *Let* $F(q,P) = P \cdot \varphi(q) + h(q)$ *with* h *in* $C^\infty(\mathbb{R}^n)$ *and* φ *a diffeomorphism ch that* $\exp(-h(q))$ *and* $\left(\frac{\partial \varphi_i}{\partial q_i}\right)$ *are of polynomial growth, i.e. they satisfy* (2.1-b). *Then*

$$T_F : S \to S$$

roof The hypothesis are chosen so that (2.1-a) can be readily verified for

$$(T\,\tilde{\Phi}(q) = \exp(-h(q))\, J(q)\, \tilde{\Phi}(\varphi(q)).$$

It is also a "well known" fact that $\hat{\tilde{\Phi}}(k)$ is in S (as a function of k of course) henever $\tilde{\Phi}$ is in . This is needed for the proof of

emma 2.3 *Let* $F(q,P) = q \cdot \varphi(P) + h(P)$ *with* $\varphi(P)$ *and* $h(P)$ *as above. Assume more- ver that* $\exp(-h(ik))$ *and* $\frac{\varphi(ik)}{i}$ *satisfy the polynomial growth conditions mentioned in* *the statement of lemma* (2.2). *Then* $T : S \to S$ *as well.*

Proof Write $F = F_1 \circ F_2$ with $F_2(q,P) = q \cdot P + h(P)$ and $F_1(q,P) = q \cdot \varphi(P)$. Accordin[g] to (1.8) we have

$$(T_{F_2}\tilde{\Phi})^\wedge(k) = \exp(-h(ik))\,\tilde{\Phi}(k) \quad (T_{F_1}\tilde{\Phi})^\wedge(k) = \hat{\tilde{\Phi}}\left(\frac{\varphi(ik)}{i}\right).$$

From this and the rest of the assumptions we obtain the desired results.

Comment From the last identities we see that when $\tilde{\Phi}(k)$ is of compact support so a[re] $(T_{F_2}\tilde{\Phi})^\wedge(k)$ and $(T_{F_1}\tilde{\Phi})^\wedge(k)$.

We shall now proceed to define the action of T_F on S' by the standard duali[ty] argument.

Definition 2.4 Let $\tilde{\Phi}$ be in S_Q and let μ be in S'_q. We put

$$\langle T_F^*\mu, \tilde{\Phi}\rangle = \langle \mu, T_F\tilde{\Phi}\rangle. \qquad (2.[...])$$

Comment We distinguish the functions of the Q-variables from by means of th[e] ~ - symbol and we add the sub-indices on S on S' to emphasize that we have functions [of] the "new" or "old" variables.

To lighten the notation we shall write T_F^* as T_F except when the risk of confusion is to[o] high.

Below we shall compute $T_F\mu$ for two classes of μ's: for μ being integration wi[th] respect to point mass (or Dirac function) and for μ denoting integration with respect to [a] measure with absolutely continuous measure with locally integrable density.

To begin with, let $\mu \in S'_q$ be of the second type, i.e., $\langle\mu,\Phi\rangle = \int f(q)\,\Phi(q)\,dq$ for Φ in S. Let $F(q,P) = P \cdot \varphi(q) + g(q)$. As we said at the beginning of the chapter, T_F is define[d] so that

$$\langle \mu, T_F\tilde{\Phi}\rangle = \int f(q)\,J(q)\exp(-g(q))\,\tilde{\Phi}(\varphi(q))\,dq$$

$$= \int f(\varphi^{-1}(Q))\exp(-g(\varphi^{-1}(Q)))\,\tilde{\Phi}(Q)\,dQ$$

nd therefore $T_F\mu$ is the functional on S_Q' consisting of integration with respect to $\tilde{f}(Q) =$ xp $- g(\varphi^{-1}(Q))\, f(\varphi^{-1}(Q))$. It is only natural that we write

$$(T_F f)(Q) = \exp(-g(\varphi^{-1}(Q))\, f(\varphi^{-1}(Q)).$$

Consider now the case where μ consists of integrating with respect to $\varepsilon_{q_0}(dq)$ or $(q - q_0)\, dq$, and $F(q,P) = P \cdot \varphi(q) + g(q)$.

$$< \mu, T_F \tilde{\Phi} > = (T_F \tilde{\Phi})(q_0) = J(q_0) \exp - g(q_0)\, \tilde{\Phi}(\varphi(q_0))$$

$$= \int \exp(-g(\varphi^{-1}(Q)))\, \delta(\varphi^{-1}(Q) - q_0)\, \tilde{\Phi}(Q)\, dQ$$

nd therefore

$$\tilde{\mu} = T_F\mu = \delta(\varphi^{-1}(Q) - q_0) \exp(-g(\varphi^{-1}(Q)))\, dQ.$$

Let now $F(q,P) = q \cdot P - h(P)$. Then for $\tilde{\Phi}(Q)$ in S_Q we know that $(T_F \tilde{\Phi})(q) =$ $\exp h(D)\, \tilde{\Phi}(q)$ and we have for the reader to verify that

$$T_F^* = \tilde{\mu} = (\exp h(-D))\mu \tag{2.6}$$

n the distribution sense. When we take $<\mu, \Phi> = \int e_k(q)\, \Phi(q)\, dq$ and $e_k(q) = \exp ik \cdot q$ ve obtain

$$< e_k, T_F \tilde{\Phi} > = \int \exp(ik \cdot q) \left(\int \exp(-iq \cdot \bar{k})\, \hat{\Phi}\left(\frac{\varphi^{-1}(\bar{k})}{i}\right) \frac{d\bar{k}}{(2\pi)^n} \right) dq$$

$$= \int \delta(\bar{k} - k)\, \hat{\tilde{\Phi}}\left(\frac{\varphi^{-1}(\bar{k})}{i}\right) d\bar{k} = \hat{\tilde{\Phi}}\left(\frac{\varphi^{-1}(ik)}{i}\right)$$

$$= \int \exp(Q \cdot \varphi^{-1}(ik))\, \tilde{\Phi}(Q)\, dQ.$$

To finish we mention a consequence of definition (2.4) containing the closest we get to nitarity with the T_F transform. For $\tilde{\Phi} \in S_Q$, $\tilde{\mu} \in S'$ and F generating a $(q,p) \to (Q,P)$ ransformation, we have

$$< \tilde{\mu}, \tilde{f} > = < \tilde{\mu}, T_{F^{-1}} T_F \tilde{f} > = < T^*_{F^{-1}} \tilde{\mu}, T_F \tilde{f} >. \quad (2.8$$

This is some sort of generalized Plancherel identity.

III.3 Evolution Semigroups Generated by Flows

Let us now look at the results of sections 1 and 2 particularized to the canonica transformations generated by the Hamilton-Jacobi functions $S(q,P,t)$ and $\Sigma(q,P,t) = S^{-1}(q,P,t)$.

Proposition 3.1 *Consider a trivially integrable system with Hamiltonian* $H(P)$. *Let* $S(q,P,t) = q \cdot P - t H(P)$ *and* $\Sigma(q,P,t) = q \cdot P + tH(P)$ *be the generating functions for th backward and forward flows respectively. Then*

(i) $F(t_2) \circ F(t_1) = F(t_1+t_2)$.

(ii) $T_{F(t_1)} T_{F(t_2)} = T_{F(t_1+t_2)}$.

Where $F(t)$ *denotes either of* $S(q,P,t)$ *or* $\Sigma(q,P,t)$.

(iii) *Defining for* Φ *such that* $\hat{\Phi}(k)$ *has compact support*

$$u(q,t) = (T_{S(t)} \Phi)(q) \qquad \bar{u}(q,t) = (T_{\Sigma(t)} \Phi)(q) \quad (3.2$$

we have

$$\frac{\partial u}{\partial t} = H(-D)\mu \qquad \frac{\partial \bar{u}}{\partial t} + H(-D)\bar{\mu} = 0; \; u(q,0) = \bar{u}(q,0) = \Phi(q). \quad (3.3$$

Proof (i) and (ii) are really easy to verify. So is (iii). To examine where the hypothese enter, note that

$$(T_{F(t)}\Phi)(q) = \int \exp(-ik \cdot q \pm H(ik)t)\, \hat{\Phi}(k) \frac{dk}{(2\pi)^n}.$$

Since $\hat{\Phi}(k)$ has compact support, the sign of t is irrelevant. Also we can differentiat under the integral sign to get

$$\frac{\partial}{\partial t}(T_{F(t)}\,\Phi)(q) = \pm\int H(ik)\exp(-ik \pm tH(ik))\,\hat{\Phi}(k)\,\frac{dk}{(2\pi)^n}$$

$$= H(-D)\int \exp(-ik\,.\,q \pm H(ik))\,\hat{\Phi}(k)\,\frac{dk}{(2\pi)^n} = \pm H(-D)\,(T_F\,\Phi)(q).$$

From this we get (3.3).

Comments

a) The awkward minus sign in $H(-D)$ is due to our definition of T_F. If we do not follow this convention some other results would look even more awkward.

b) Notice that $S(q,P,t)$ ($\Sigma(q,P,t)$ respectively) maps the coordinates at present time t to the coordinates at $t = 0$ (the coordinates at time 0 to coordinates at time t respectively). So in $T_{S(t)}\Phi(q)$ ($T_{\Sigma(t)}\Phi(q)$ respectively) q is the "present coordinate" of the system (the "old coordinate" of the system, respectively) and $\frac{\partial u}{\partial t} = H(-P)u$ and $\left(\frac{\partial \bar{u}}{\partial t} + \frac{H(-D)}{\bar{u}}\right) = 0$ are forward and backward equations, respectively.

Proposition 3.4 *Let* $H(p)$ *and* $\Phi(\cdot)$ *be as in proposition* (3.1). *Let* $\langle \mu, \Phi \rangle$ *be a given functional, and if we define* $\mu(t)$ *by* $\langle T_f\mu, \Phi \rangle \equiv \langle \mu(t), \Phi \rangle = \langle \mu, T_F\Phi \rangle$ *with* F *as in proposition* (3.1). *Then*

$$\frac{\partial \mu(t)}{\partial t} \pm H(-D)\mu = 0, \qquad \mu(0) = \mu \tag{3.5}$$

in the distribution sense. The minus or plus signs are to correspond with $F = S$ *or* $F = \Sigma$ *respectively.*

Proof Just differentiate the defining relationship with respect to t.

From section (II.4) we know that all integrable systems on $\mathbb{R}^n$ are canonically equivalent. From (3.1) we know how to associate (semi) groups to them. So it is natural to have

Proposition 3.6 *Let* $H(p)$ *and* $\tilde{H}(p)$ *be two (real analytic) Hamiltonians on* $\mathbb{R}^n$. *Let* $u(q,t)$ *and* $\tilde{\mu}(Q,t)$ *be given by* (3.2) for $S(q,\bar{p},t) = q\,.\,\bar{p} - tH(\bar{p})$ *and* $\tilde{S}(Q,\bar{P},t) = Q\,.\,\bar{P} - \tilde{H}(\bar{P})$. *Let* $\tilde{\Sigma}$ *be the inverse of* $\tilde{S}$. *Then if* $F(q,P,t) = \tilde{\Sigma}(t) \circ S(t)(q,p) = q\,.\,P + \tilde{H}(P) - tH(p)$ *we have*

$$\mu(q,t) = (T_F\tilde{\mu})(q,t). \tag{3.7}$$

Proof One proof is the following. Since $T_F = T_{\tilde{\Sigma} \circ S} = T_S T_{\tilde{\Sigma}}$ we have

$$T_F \tilde{u} = T_S T_{\tilde{\Sigma}} T_{\tilde{S}} = T_S \Phi = \mu$$

since $T_{\tilde{\Sigma}} = T_{\tilde{S}}^{-1}$

The other proof consists in taking Fourier transforms of (3.7) and verify it using (1.8) since

$$\hat{u}(k,t) = \exp tH(ik)\, \hat{\varphi}(k) = \exp t(\tilde{H}(ik) - \tilde{H}(ik)) \exp t\tilde{H}(ik)\hat{\Phi}(k)$$

$$= (T_F \tilde{u})^{\wedge}(k).$$

Comments

(a) An analogous result can be obtained for the m(t)'s defined in proposition (3.4).

(b) The intuitive idea behind the proof is contained in the physical meaning of the canonical transformation generated by $F = \tilde{\Sigma} \circ S$: first go back to time $t = 0$ along the flow generated by H and then go forward to present time t along the flow generated by $\tilde{H}$.

(c) Processes with independent increments are naturally associated to integrable systems. Proposition (3.6) provides a map between the transition semigroups of the involved processes. The big thing would be to carry the mapping onto the path spaces of each process.

Proposition 3.8 *Let* $S(t)$ *and* $\tilde{S}(t)$ *be as above and let now* $F(q,P) = q \cdot U(P)$ *be the generating function of a time independent canonical transformation such that* $\tilde{\tilde{\Phi}}(P) = H(U(P))$. *Let* $\tilde{\Phi}(Q)$ *and* μ *in* S_q' *be given and let* $\Phi = T_F\Phi$ *and* $\tilde{\mu} = T_F\mu$, *then, if we put* $P_t = T_{S(t)}$, $\tilde{P}_t = T_{\tilde{S}(t)}$ *we have*

$$T_F \tilde{P}_t \tilde{\Phi}(q) = P_t \tilde{T}_F \tilde{\Phi}(q) \tag{3.9}$$

and also

$$\langle T_F^{-1} \tilde{\mu}, P_t T_F \tilde{\Phi} \rangle = \langle \tilde{\mu}, \tilde{P}_t \tilde{\Phi} \rangle. \tag{3.10}$$

roof Assume (3.9). Applying the definitions we obtain

$$<T_F^{-1}\tilde{\mu}, P_t T_F \tilde{\Phi}> = <T_F^{-1}\tilde{\mu}, T_F \tilde{P}_t \tilde{\Phi}>$$

$$= <\tilde{\mu}, T_F^{-1} T_F \tilde{P}_t^{\tilde{}}> - <\tilde{\mu}, \tilde{P}_t \tilde{\varphi}>$$

hich is (3.10). A variant form of (3.10) is given by

$$<\tilde{\mu}, \tilde{P}_t \tilde{\Phi}> = <T_F \mu, \tilde{P}_t \tilde{\Phi}> = <\mu, T_F \tilde{P}_t T_F^{-1} \Phi> = <\mu, P_t \Phi>.$$

To obtain (3.9) it actually suffices to consider the Fourier transforms of both sides of the lentity and verify that they coincide. Begin with the left hand side. According to (1.8)

$$(T_F \tilde{P}_t \tilde{\Phi})^{\wedge}(k) = (\tilde{P}_t \tilde{\Phi})^{\wedge}\left(\frac{V(ik)}{i}\right) = \exp t\tilde{H}(V(ik))\, \hat{\tilde{\Phi}}\left(\frac{V(ik)}{i}\right).$$

Consider now the right hand side of (3.9)

$$(P_t T_f \tilde{\Phi})^{\wedge}(k) = \exp tH(i\, \tilde{\Phi}(i\ p\ tH(ik)\, \hat{\tilde{\Phi}}\left(\frac{V(ik)}{i}\right)$$

here in both cases repeated use of the two identities in (1.8) was made of. The result now rops out since $\tilde{H}(P) = H(U(P))$. This completes the proof of (3.8).

omment The moral in proposition (3.8) is that we have to transform initial conditions roperly before applying a transformed semigroup. Another way to look at (3.9) is the ollowing. If you vary U over the class of diffeomorphisms on $\mathbb{R}^n$, we obtain an orbit f Hamiltonians $\tilde{H}_U = H(U)$. Then the class of semigroups they generate is given by $= T_F^{-1} P_t T_f = T_{F^{-1}} P_t T_F$.

An important variation on the same theme is contained in

emma 3.11 *Let* μ_t *be a convolution semigroup of measures with Fourier transforms*

$$\int \mu_t(dq) \exp(-ik\,.\,q) \equiv \hat{\mu}(k) = \exp tH(ik)$$

or an appropriate H, *analytic on* $\mathbb{R}^n$. *Let* $F(q,P) = q\,.\,U(P)$ *be as in proposition* (3.8) *nd we put* $\hat{\mu}_t = T_F^* \mu_t$, *then*

$$\hat{\mu}_t(t) = \exp t\tilde{H}(ik) = \exp H(U(-ik)).$$

Proof Notice to begin with that

$$< \mu_t, \Phi > = \int \mu_t(dq)\, \Phi(q)\, dq = \int \mu_t(-k)\, \hat{\Phi}(k) \frac{dk}{(2\pi)^n}$$

$$= \int \exp(tH(-ik))\, \hat{\Phi}(k) \frac{dk}{(2\pi)^n} = < \varepsilon_0, T_{F_R(t)}\, \Phi > = < T^*_{F_R(t)}\, \varepsilon_0, \Phi >$$

where $S = q \,.\, P - t\, H(P)$. Note as well that for any test function $\tilde{\Phi}(Q)$

$$< \varepsilon_0, T_F\, \tilde{\Phi} > = < \varepsilon_0, \tilde{\Phi} > = < T^*_F \varepsilon_0, \tilde{\Phi} >.$$

Consider now, for a test function $\tilde{\Phi}(Q)$

$$< \varepsilon_0, T_{S(t)}\, T_F\, \tilde{\Phi} > = < \varepsilon_0, T_F\, T_{\tilde{S}_{(t)}}\, \tilde{\Phi}) = < T^*_{\tilde{S}(t)}\, T^*_F\, \varepsilon_0, \tilde{\Phi} >.$$

By dualization, from $T_F\, T_{\tilde{S}(t)} = T_{S(t)}\, T_F$ we obtain $T_F\, T_{\tilde{S}(t)} = T_{S_R(t)}\, T_F$, we obtain

$$T^*_{\tilde{S}(t)}\, T^*_F = T^*_F = T^*_{S(t)}$$

therefore

$$T^*_F\, T^*_{S(t)} \varepsilon_0 = T^*_F \mu_t = T^*_{\tilde{S}}\, T^*_F\, \varepsilon_0 = T^*_{\tilde{S}_R(t)}\, \varepsilon_0 = \tilde{\mu}_t.$$

Rewriting

$$< \varepsilon_0, T_{S(t)}\, T_F\, \tilde{\Phi} > = < \varepsilon_0, T_{\tilde{S}(t)}\, \tilde{\Phi} > = < \tilde{\mu}_t, \tilde{\Phi} >$$

in terms of Fourier transforms we obtain

$$\int \hat{u}_t(k)\, \hat{\tilde{\Phi}}(k) \frac{dk}{(2\pi)^n} = \int \exp(t\, \tilde{H}(ik))\, \hat{\tilde{\Phi}}(k) \frac{dk}{(2\pi)^n}$$

ith $\tilde{H}(P) = H(U(P))$. Therefore

$$\hat{\tilde{\mu}}_t(k) = \exp tH(- U(- ik)$$

which concludes our proof.

4.4 The General Case

The correspondence between the Hamiltonian flows of trivially integrable systems and evolution semigroups carried out on the previous section hinged on the fact that $T_{F(t)}$ turned out to be a (semi)group as mentioned in (3.1)-(ii).

The obvious natural question is then: can the program of section 3 be carried out for systems with arbitrary Hamiltonian $H(q,p)$? The first thing to do would be to solve $\frac{\partial S}{\partial t} + H(q,\nabla S) = 0$, $S(q,P,0) = q \cdot P$ and then to try to define $T_{S(t)}$.

We know how to obtain $S(q,P,t)$ from its infinitesimal components. This was carried out in section II.3. In the previous sections we saw how to represent those infinitesimal components by appropriate integral operators. Let us try to combine both ingredients together.

Recall from (II-3) that

$$S^{(N)}(q^0,p^N,t) = S_N^{(N-1)} \circ S_N^{(N-2)} \circ \ldots \circ S_N^{(0)}(q^0,p^N,t)$$

tends to $S(q,P,t)$ as N tends to infinity.

The $S_N^{(k)}(q^{k-1},p^k,t)$ were given by $q^{k-1} \cdot p - \frac{tH(q^{k-1},p^k)}{N}$ which when $H(q,p) = K(p) + V(q)$ can be written as $S_N^k(q^{k-1},p^n) = F_K(\varepsilon) \circ F_V(\varepsilon)$ with $\varepsilon = \frac{t}{N}$ and

$$(q^k,p^k) = \left(q\left(t - \frac{kt}{N}\right), p\left(t - \frac{kt}{N}\right)\right).$$

According to what we know we have

$$T_{S_N(t)} = T_{S_N^{(0)}} T_{S_N^{(1)}} \ldots T_{S_N^{(N-1)}}$$

and furthermore every individual factor is given by

$$T_{S_N^k(\varepsilon)} = T_{F_K(\varepsilon) \circ F_v(\varepsilon)} = T_{F_v(\varepsilon)} T_{F_K(\varepsilon)}.$$

To simplify notations a bit we shall put

$$U_k(t) = T_{F_K(\varepsilon)}, \qquad U_v(t) = T_{F_v(t)} \cdot$$

We have to examine the limits

$$\lim_{N\to\infty} T_{S^N} = \lim \prod_0^{N-1} T_{S_N^{(k)}} = \lim_{N\to\infty} \left(U_v\left(\frac{t}{N}\right) U_K\left(\frac{t}{N}\right)\right)^N \tag{4.2}$$

which have been scrutinized quite a bit in a variety of contexts under the names of Lie Trotter-Kato-Chernoff product formulas.

In chapter V we shall rewrite (4.2) in terms of expectation values associated to the fre pa K(p) of H(q,p) and we shall obtain a variation on the theme of the Feynman-Kac formula.

The problem here being that even if $T_{S(t)}$ were well defined, it may not turn out to be semigroup. The reason why this does not happen when considering trivially integrabl systems was explained in the conlcuding paragraphs of chapter II.

To conclude we have

Proposition 4.3 *Let* S(t) *be the Hamilton-Jacobi function corresponding to a Hamiltonia* H(q,p). *Assume that* $u = T_{S(t)}\Phi$ *is well defined by* (1.3) *and that* $T_{S(t+u)} = T_{S(t)} T_{S(u)}$. *The*

$$\frac{\partial u}{\partial s} = G(q,D)u \qquad u(q,0) = f(q) \tag{4.4}$$

with G(q,D) = H(q,D) + δ(q,D) *where the* D *operators are to be put on the right of the* q' *Here* δ(q,p) *is given by*

$$\delta(q,P) = \frac{\partial}{\partial t} \det \left(\frac{\partial^2 S}{\partial q_i \partial P_j} \right)\Bigg|_{t=0} .$$

Proof Start from

$$u(q,t) = (T_{S(t)}\, \Phi)(q) = \int \Delta(q,ik,t) \exp\left(-S(q,ik,t)\right) \tilde{\Phi}(k) \frac{dk}{(2\pi)^n}.$$

Now differentiate with respect to t and compute at t = 0, use S(q,P,0) = q . P to obtain

$$\left.\frac{\partial u}{\partial s}\right|_{s=0} = \int \left(H(q,ik) + \delta(q,ik)\right) \exp\left(-ikq\right) \hat{\Phi}(k) \frac{dk}{(2\pi)^n} \tag{4.5}$$

$$= (G\, \Phi)(q)$$

The comment about ordering is related to the integral representation. Since T is assumed to have the semigroup property, (4.4) follows from (4.5).

III.5 Analogies with Quantum Mechanics

Quantum mechanics is part of the mathematical description of "physical reality" with a long list of impressive successes. Most of the "intellectual fun" it provides comes not only from the picture of the world we get from it, but from the sense of flow and controversy in its very foundations.

There are several presentations of quantum mehcanics, but here we very briefly sketch the standard one. The states of any system are assumed to be vector in a Hilbert space $\mathbb{H}$. The physical observables of the system are described by self-adjoint operators on (dense domains in) $\mathbb{H}$ and the evolution of the system is to be given by a one-parameter group $U(t)$ of unitary operators on $\mathbb{H}$.

To specify a system with a classical analogue, i.e., a system of interacting particles, a set of correspondence rules is established for a particle moving in $\mathbb{R}^n$. The space $\mathbb{H}$ is taken to be $L^2(\mathbb{R}^n)$, and to the q_i-th coordinate we assign the operator $\hat{q}_i$ of multiplication by q_i and to the p_i-th component of the momentum vector we assign the operator $p_i = -i\dfrac{\partial}{\partial q_i}$. (I am using the convention which assigns the value 2π to Planck's constant h).

Then comes the problem of how to assign operators corresponding to classical observables which are functions $f(q,p)$ and, after that one is solved, to compute quantities like

$$\langle A \rangle_{\psi}(t) = \langle \psi(t) | A / \beta(t) \rangle \,|\, \langle \psi(t)/\psi(t) \rangle \tag{5.1}$$

which are interpreted as the mean value of the observable A in state $\langle \psi(t) \rangle \equiv \langle U(t)/\psi) \rangle$. Here $\langle \varphi|\psi \rangle$ denotes the scalar product in $L^2(\mathbb{R}^n)$.

A quantum analogue of the classical equivalence of trivially integrable systems is contained in

Theorem 5.2 *Let* $U_1(t)$ *and* $U_2(t)$ *be the time evolution operators of two quantum systems with the same state space* $\mathbb{H}$. *There exists a time dependent family* $W(t)$ *of unitary operators such that if* $\psi(t)$ *satisfies* $i\dfrac{\partial\psi}{\partial t} = H_1\,\psi(t)$ *(or* $\psi(t) = U_1(t)\,\psi(0) = \exp - itH_1\,\psi(0)$*) then* $\tilde{\psi}(t) = W(t)\,\psi(t)$ *satisfies* $i\dfrac{\partial\tilde{\psi}}{\partial t} = H_2\,\tilde{\psi}$ *(or* $\tilde{\psi}(t) = U_2(t)\,\psi(0) = \exp - it\,H_2\,\psi(0)$*).*

Proof The obvious candidate is $W(t) = U_2(t)\,U_1(-t)$.

Comments

(i) This is the quantum analogue of the Lie–Koenig theorem. See reference [6] of chapter II.

(ii) The obvious question is: what $W(t)$ do come from unitary representations of canonical transformations?

(iii) In [11] an application of these ideas is presented.

We should comment upon another parallel of what we do here to the problem of quantization of a mechanical system. One of the ways to tackle the issue of estalbishing correspondence rules comes under the name of geometric quantization. We refer the reader to [6] - [7]. In [6] a summary description is given and applications to quantum optics are presented.

The idea consists in looking for unitary representations of the group of canonical transformations, which leaves the canonical commutation relations

$$[q_i, p_j] = \delta_{ij} \tag{5.3}$$

invariant.

The bracket in (5.3) is the Poisson bracket defined for any two functions $f(q,p)$, $g(q,p)$ by

$$[f, g] = \sum \frac{\partial f}{\partial q_i}\frac{\partial g}{\partial p_i} - \frac{\partial f}{\partial p_i}\frac{\partial g}{\partial q_i}.$$

One has to realize that any $f(q,p)$ can be taken as the infinitesimal generator of a one-parameter gorup of canonical transformations. With the aid of the appropriate unitary representation a one-parameter group of unitary transformations is obtained. In particular the Hamiltonian operator in quantum mechanics wil be the infinitesimal generator of the group $U(t)$ by unitarily representing the canonical group of transformations generated by the classical Hamiltonian operator. This group turns out to be the flow T_t mentioned several times above.

References

In this chapter we hope to have expanded, explained and corrected

[III-1] Gzyl, H.: "Evolution semigroups and Hamiltonian flows". Jour. Math. Analysis and Appl. Vol 110, (1985) 316-332.

[III-2] Gzyl, H. : "The Feynman-Kac Formula and the Hamilton-Jacobi Equation". Jour-Math. Analysis and Appl. (To appear).

Evolution equations related to integrable hamiltonians have been considered by

[III-3] Brode, Johm : "Solutions to the Cauchy Problem for the Family of Stable Probability Laws".

[III-4] Berger, M. A. and Sloan, A. D. : "A Method of Generalized Characteristics". Memoirs of the A. A. S. Vol. 38, 1982.

An introduction to semigroups and interesting examples is

[III-5] Goldstein, J. A. : "Semigroups of Linear Operators and Applications". Oxford Univ. Press. New York, 1985.

The problem of geometric quantization can be seen in

[III-6] Raszillier, H and Schempp, W. : "Fourier Optics from the Perspective of the Heisenberg Group".

[III-7] Hurt, N. : "Geometric Quantization in Action". D. Reidel Publishing Co. Dordrecht (1983).

From the following much of the references to previous work on unitary representations can be chased

[III-8] Moshinsky, M. and Quesne, C. : "Linear Canonical Transformations and their Unitary Representations". J. Math. Phys. Vol 12, No. 3, (1971) 1772 - 1783.

[III-9] Kramer, P., Moshinski, M. and Seligman, T. H. : "Non-bijective Canonical Transformations and their Representations in Quantum Mechanics" J. Math. Phys. Vol. 19, No.3 (1978) 683 - 693.

[III-10] Wolf, K. B. "The Heisenberg-Weyl Ring in Quantum Mechanics" in "Group Theory and its Applic.". Vol III. Ed. Loebl, E. M., Acad. Press, N. Y. (1975).

To finish, we mentioned in section 5

[III-11] Gzyl, H. : "Bounded v.s. Free". Hadronic Jour. Vol. 8, No. 2 (1985) 119 - 120.

Chaper IV
Operational Calculus

IV.1 The Correspondence Rules

This chapter is concerned with that aspect of operational calculus that consists on obtaining formulas and formal identities between differential operators without paying attention to questions about domains of definition, topological properties, etc. We provide some references in which these matters are properly dealt with. Here we develop enough material to extend Feinsilver's results [1-2] to more general cases. This is carried out in the next two chapters.

In some of the computations above we associated an operator $f(-D)$ to a function $f(p)$. Two basic ways in which this can be done are the following. We can set

$$f(-D) = \int \hat{f}(k) \exp(+ik \cdot D) \frac{dk}{(2\pi)^n} \tag{1.1}$$

or by setting, when $f(p) = \sum f_m p^m / m!$

$$f(-D) = \sum f_m (-D)^m / m! \tag{1.2}$$

Certainly (1.2) can be obtained from (1.1) by expanding the exponential and setting $\int (-ik^m \hat{f}(k) \frac{dk}{(2\pi)^n}$ equal to f_m.

In order to associate differential operators to functions $f(q,p)$ we follow the procedures dictated by the theory of representations of canonical transformations introduced in chapter III, combined with a way of generating canonical transformation by an extension of the Hamilton-Jacobi theory.

So given a function $f(q,p)$ we consider the group of canonical transformations generated by $W(q,P,s)$ satisfying

$$\frac{\partial W}{\partial s} + f(q,\nabla,W) = 0 \qquad W(q,P,0) = q \cdot P \tag{1.3}$$

and consider the operators

$$(T_{W(s)}\,\Phi)(q) = \int \Delta(s)\exp(-\,W(q,ik,s))\,\hat{\Phi}(k)\,\frac{dk}{(2\pi)^n}. \tag{1.4}$$

Now we take the derivative of (1.4) with respect to s and evaluate at $s = 0$ to obtain

$$\partial T_{W(s)}\,\Phi(\partial s(q))\,|_{s=0} = \int G(q,ik)\exp(-ik\,.\,\mathbf{q})\,\hat{\mathbf{\Phi}}(k)\,\frac{dk}{(2\pi)^n}$$

where $G(q,ik) = f(q,ik) + \delta(q,ik)$, with $\delta(q,p)$ being as in the proof of proposition (III–4.4). Se below anyway

Definition 1.5 *To a smooth function* f(q,p) *we associate a differential operator, denoted by* f(q,p) *as follows*

$$(f_{op}(q,p))\,\Phi(q) \tag{1.6-1}$$

$$= \int [\,f(q,ik) + \delta(q,ik)\,]\exp(-ik\,.\,q)\,\hat{\Phi}(k)\,\frac{dk}{(2\pi)^n}$$

with

$$\delta(q,p) = \left.\frac{\partial \det|\,\partial^2 W/\partial q_i\,\partial P_j\,|}{\partial s}\right|_{s=0} \tag{1.6-2}$$

Comments Notice that when $\delta(q,p) = 0$, then (1.6–1) gives us the obvious answer. (Clearly seen when f(q,p) is a power series in $q^n\,p^m$). Just write the q's to the left of the p's and replace each p_i by $\frac{-\partial}{\partial q_i}$.

Let us now work out some obvious cases. We begin by considering $f = f(q)$ only. In this case (1.3) is just $\frac{\partial W}{\partial s} + f(q) = 0$ and $W(q,P,s) = q\,.\,P - s\,f(q)$. In this case $\Delta \equiv 1$ and

$$(T_{W(s)}\,\Phi)(q) = \exp s\,f(s)\,\Phi(q)$$

which yields according to (1.6)

$$(f_{op}(q))\,\Phi(q) = f(q)\,\Phi(q) \tag{1.7}$$

which is the expected result. Let us now consider a variation on the theme of (1.1). Let $= f(p)$ only. In this case $\frac{\partial W}{\partial s} + f(\nabla W) = 0$, $W(q,P,0) = q\,.\,P$ is satisfied by the obvious $W(q,P,s) = q\,.\,P - f(P)\,s$. In this case $\Delta \equiv 1$ as well and

$$(T_{W(s)})\,\Phi(q) = \int \exp(-iq\,.\,k + f(ik)\,s)\,\hat{\Gamma}(k)\,\frac{dk}{(2\pi)^n}$$

from which we obtain (no surprise this time)

$$(f_{op}(p)\,\Phi(q) = \int f(ik)\exp(-ik\,.\,q)\,\hat{\Phi}(k)\,\frac{dk}{(2\pi)^n} = f(-D)\,\Phi(q)$$

which is (1.1) again. A particular case of this result, that corresponding to $f(p) = p_i$ shows how the representation formula (III-1.3) is responsible for the minus sign in the correspondence $p_i \to \frac{-\partial}{\partial q_i}$. For $f(p) = p_i$, $W(s) = q\,.\,P - s\,p_i$ and

$$(T_{W(s)}\,\Phi)(q) = \Phi(q - s\,p_i)$$

from which we see that $\left.\frac{\partial T_{W(s)}\Phi(q)}{\partial s}\right|_{s=0} = ((f_i)_{op}\,\Phi)(q) = -\frac{\partial \Phi}{\partial q_i}$.

To close this section let us consider the case in which $f(p,q) = q\cdot p$. Now the equation for $W(s)$ becomes $\frac{\partial W}{\partial s} + q\,.\,\frac{\partial W}{\partial q} = 0$ with $W(q,P,0) = q\,.\,P$.

The Hamiltonian system associated to $f(q,p)$ is (see (II-6.4)) $\frac{dq}{ds} = q$, $\frac{dp}{ds} = -p$ which is readily solved. From this and (3.12) we obtain $q = q_0\exp(s)$, $p = p_0\exp(-s)$

$$W(q,P,s) = P\,.\,q\exp(-s) + \int_0^t (p(s)\,.\,\dot{q}(s) - q(s)\,.\,p(s)]\,ds$$

$$= P\,.\exp(-s)$$

where we replace p_0 by P to conform with the notation we are using. Now $\Delta = \exp(-s)$ and therefore $\delta = -n$, and so

$$((pq)_{op}\,\Phi) = \int [\,q\cdot(ik) - n\,]\exp(-kq)\,\hat{\Phi}(k)\,\frac{dk}{(2\pi)^n}$$

$$= -\,(q\,D + 1)\,\Phi(q) = -\,D\,q\,\Phi(q).$$

To cook up a 1-dimensional example in which $\delta(q,p)$ is not constant, let us proceed as follows. Let $f(q,p) = g(q)\,p$, with $g(q) > 0$ such that $\frac{dG(q)}{dq} = \frac{1}{g(q)}$ for an appropriate invertible $G(q)$. Let $R(\xi)$ be the inverse of G. In this case the Hamiltonian system associated to $f(q,p) = g(q)\,p$ is

$$\dot{q} = g(q); \qquad \dot{p} = -\,p\,\frac{d}{dq}g(q)$$

which can be integrated to $q(s) = R\,(G(Q) + s)$ and $\frac{P(s)}{P} = \frac{g(0)}{g(q)}$ where the initial (final for the transformation generated by $W(s)$) state is denoted by (Q,P). If we compute the action according to (3.12) we obtain

$$W(q,P,s) = P\,R\,(G(q) - s)$$

which is what we would have obtained had we applied the method of characteristics to

$\frac{\partial W}{\partial s} + g(q)\,\frac{\partial W}{\partial q} = 0$, $W(q,P,0) = q\,.\,P$. Here

$$\Delta = \frac{dR}{d\xi}(G(q) - s) \quad \text{and} \quad \frac{d^2R(G(q))}{d\xi^2} = g(q)\,\frac{dg}{dq}(q) = \delta(q).$$

In this case

$$(f(q,p)_{op}\,\Phi) = [\,-\,g(q)\,D + \delta(q)\,]\,\Phi(q).$$

Observe now that if $X(q)$ is any locally integrable function, we can define functions of operators acting on $X(q)$ by the appropriate dualization. We let $X(q)$ act on test functions by

$$\langle X, \Phi \rangle = \int X(q)\,\Phi(q)\,dq$$

and define

$$\langle f_{op}\,X, \Phi \rangle \equiv \langle X, f_{op}\,\Phi \rangle. \tag{1.9}$$

The simple cases treated above now become

i) When $f = f(q)$ only, then

$$< X, f_{op} \Phi > = \int X(q) [f(q) \Phi(q)] dq = \int [f(q) X(q)] \Phi(q) dq$$

from which we obtain

$$f_{op}(q) X(q) = f(q) X(q).$$

ii) When $f = f(p)$ we obtain

$$< X, f_{op} \Phi > = \int X(q) f(-D) \Phi(q) dq = \int [f(D) X(q)] \Phi(q) dq$$

or as it should be

$$[f_{op}(q) X](q) = f(D) X(q).$$

iii) To finish let $f(q,P) = q\,p$. From what we saw above

$$< X, f_{op} \Phi > = \int X(q) [- D q \Phi(q)] dq = \int [q D X(q)] \Phi(q) dq$$

or

$$(q\,p)_{op} X(q) = q D X(q).$$

IV.2 The Generalized Leibnitz Lemma and other Formulae

Let us now obtain an extension of the standard Leibnitz lemma. Since our procedure allows for the full generality we shall do so. We shall associate to functions $f(q,p)$ a different class of differential operators defined by

$$(\hat{f}_{op} \Phi(q) = \int f(q,ik) \exp(-ik \cdot q) \hat{\Phi}(k) \frac{dk}{(2\pi)^n} \tag{2.1}$$

and we start by computing the Fourier transform of this object.

$$\int \exp(ik \cdot q) (\hat{f}_{op} \Phi^\wedge)(q) dq$$

$$= \int \exp(ik \cdot q) \int f(q,i\bar{k}) \exp(-i\bar{k} \cdot q) \hat{\Phi}(\bar{k}) \frac{d\bar{k}}{(2\pi)^n}$$

$$= \int dk' \int \hat{f}(k', ik)\, \delta(k - k' - \bar{k})\, \hat{\Phi}(\bar{k}) \frac{d\bar{k}}{(2\pi)^n}$$

where the missing steps follow from

$$\int \exp(-ik \cdot q)\, h_1(q)\, h_2(q)\, dq$$

$$= \int \hat{h}_1(k')\, \hat{h}_2(k - k') \frac{d\bar{k}}{(2\pi)^n}$$

for any $h_1(k)$ and $h_2(k)$ and the δ-function representation (III-1.1). Integrating over k' above we obtain

$$\int \exp(ik \cdot q)\, [\hat{f}_{op}(\,)(q)]\, dq = \int \hat{f}(k - \bar{k}, i\bar{k})\, \hat{\Phi}(\bar{k}) \frac{d\bar{k}}{(2\pi)^n}. \tag{2.2}$$

We can proceed now to

Proposition 2.2 (GLL) *For smooth* f,g *having convergent Taylor expansions, to which* $\hat{f}_{op}$ *and* $\hat{g}_{op}$ *can be associated as above and for any test function* Φ, *we have*

$$\hat{f}_{op}\, \hat{g}_{op}\, \Phi(q) = \sum \frac{1}{m} \int (D_q^m g)(q,ik)\, (D_p)^m f(q,ik)\, \hat{\Phi}(k) \frac{dk}{(2\pi)^n}. \tag{2.3}$$

Corollary 2.4 *For* $g = g(q)$ *and* $f = f(p)$ *we obtain*

$$f(-D)\, g(q)\, \Phi(q) = \sum \frac{1}{m!} (D^m g)(q)\, (-D^m f)(-D)\, \Phi(q). \tag{2.5}$$

Proof Apply the previous proposition.

Proof of 2.3 According to (2.1) and (2.2)

$$\hat{f}_{op}\, \hat{g}_{op}\, \Phi(q)$$

$$= \int f(q,ik) \exp(-ik \cdot q) \left(\int \hat{g}(k - \bar{k}, i\bar{k})\, \hat{\Phi}(\bar{k}) \frac{d\bar{k}}{(2\pi)^n} \right) \frac{dk}{(2\pi)^n}$$

$$= \int f(q,i(k+\bar{k}))\exp(-i(k+\bar{k})\,.\,q)\left(\int \hat{g}(k,i\bar{k})\,\hat{\Phi}(\bar{k})\frac{d\bar{k}}{(2\pi)^n}\right)\frac{dk}{(2\pi)^n}$$

$$= \sum_m \frac{1}{m!}\int (D_p^m f)(q,i\bar{k})\left\{\int (-ik)^m\,\hat{g}(k,i\bar{k})\exp(-ik\,.\,q)\frac{dk}{(2\pi)^n}\right\}$$

$$\exp(-i\bar{k}\,.\,q)\,\hat{\Phi}(\bar{k})\frac{dk}{(2\pi)^n}$$

$$= \sum_m \frac{1}{m!}\int (D_p^m f)(q,i\bar{k})\,(-D_q)^m g(q,i\bar{k})\exp(-i\bar{k}\,.\,q)\,\Phi(\bar{k})\frac{d\bar{k}}{(2\pi)^n}$$

$$= \sum_m \frac{1}{m!}\int (D_p^m g)(q,ik)\,[(-D_p)^m f\,](q,ik)\,\hat{\Phi}(k)\frac{dk}{(2\pi)^n}$$

which is what we want and which differs from the previous expression only by relabelling and rearranging the terms.

Corollary 2.6 *With the notations introduced above we have*

$$[\hat{f}_{op}, \hat{g}_{op}]\,\Phi(q)$$

$$= \sum_m \frac{1}{m!}\int \{\,D_p^m g)(q,ik)\,(-D_p^m f)(q,ik)$$

$$-(D_p^m f)(q,ik)\,(-D_p^m g)(q,ik)\,\}\exp(-ik\,.\,q)\,\Phi(k)\frac{dk}{(2\pi)^n}.$$

Comment *The order under the integral sign in* (2.5). The order on the left hand side determines which function gets differentiated with respect to which variable.

IV.3 Some Exponential Formulae

In a few places in chapters V and VI we shall need the results of the computations we present below. They are a variation on the Lie-Hausdorff-Baker-Campell-etc. product formulae.

We shall be making extensive use of the following two rather common identities

$$\exp(f(D))\exp(a\,.\,q) = (\exp f(a))\,(\exp(a\,.\,q) \qquad (3.1\text{-a})$$

$$\exp(b\,.\,D)\,\Phi(q) = \Phi(q+b) \tag{3.1-b}$$

with f such that f(D) is defined and a,b are vectors, etc.

Proposition 3.3 *Let* K(p) *such that* $X(D) = (\nabla K)(-D)$ *is defined. For* $b \neq 0$, a, l, k in $\mathbb{R}^n$ *and,* $e_k = \exp(-ik\,.\,q)$ *we have*

$$\exp\{l\,.\,(a\,X(D) + bq)\}e_k(q)$$
$$= \exp\frac{a}{b}\{K(ik) - K(ik - bl)\}\exp q\,.\,(bl - ik). \tag{3.4}$$

Proof We shall verify (3.4) using

$$\exp l\,.\,(a\,X(D) + bq)\,e_k(q)$$

$$= \lim_n\left(\exp\frac{al\,.X(D)}{n}\exp\frac{bl\,.\,q}{n}\right)^n e_k(q).$$

To begin with, observe that according to (3.1-a)

$$\exp\frac{l\,.X(D)}{n}\exp q\,.\left(\frac{bl}{n} - ik\right)$$

$$= \exp al\,.\,X\left(\frac{bl}{u} - ik\right)\exp q\,.\left(\frac{bl}{n} - ik\right)$$

which after n iterations yields

$$\left(\exp\frac{al\,.X(D)}{n}\exp\frac{bl\,.\,q}{n}\right)^n e_k(q).$$

$$= \exp\sum_1^n\frac{al}{n}.\,X\left(-ik + \frac{bl}{n}j\right)\exp q\,.\,(bl - ik).$$

It is clear that the limit process can be carried out on the exponential on the right side. The limit exponent is

$$\lim_{n\to\infty} \sum_{1}^{n} \frac{al}{n} . X\left(-ik + \frac{bl}{n} j\right) = \frac{a}{b} \int_{0}^{1} bl . \nabla K(-ik - bls)\, ds$$

$$= \frac{a}{b} \{K(ik) - K(ik - bl)\}$$

rom which (3.4) drops out.

Comments When b tends to 0, both sides of (3.4) tends to the same value. When $K(p) = \frac{p^2}{2}$ we obtain

$$\exp l . (bq - aD)\, e_k(q) = \exp a\left(ik . l - \frac{bl^2}{2}\right) \exp q . (bl - ik). \tag{3.5}$$

The complementary case is contained in

Proposition 3.6 *Let* U(q) *be continuously differentiable on* $\mathbb{R}^n$ *and put* $F(q) = \nabla U(q)$. *For* l,k *in* $\mathbb{R}^n$; $b, a \neq 0$, *in* $\mathbb{R}$ *we have*

$$\exp l . ((aD + bF(q))\, e_k(q)$$

$$= \exp ial . k \exp \frac{b}{a}\{ U(q + la) - U(q) \}\, e_k(q). \tag{3.7}$$

Proof Now proceed as follows

$$\exp l . (aD + bF(q))\, e_k(q)$$

$$= \lim_{n} \left(\exp \frac{al . D}{n} \exp \frac{bl . F(q)}{n}\right)^n \tag{3.7}$$

and combine it with (3.1-b) to obtain

$$\left(\exp \frac{al . D}{n} \exp \frac{bl . F(q)}{n}\right) e_k$$

$$= \exp\left\{ \sum_{1}^{n} \frac{bl}{n} \cdot F(q+\frac{al}{n}j) - ik\,.\,(q+al) \right\}$$

which converges to (2.7) as n tends to infinity.

Corollary 3.8 *Let* $U(q) = (q,Aq)/2$ *with* A *being a symmetric* $n \times n$*-matrix. For* $F(q) = Aq$, (3.7) *yields*

$$\begin{aligned} &\exp l\,.\,(aD + bAq) \\ &= \exp\{b(l\,.\,Aq) + ab(l\,.\,Al)/2\}\exp - ik\,.\,(q+al) \end{aligned} \tag{3.9}$$

which coincides with (3.7) *after we change* a *into* $-a$.

IV.4 Transformation of Operators under Canonical Transformations

In chapter III we saw how to transform functions of coordinates under the action of canonical transformations. Above we saw how to associate operators to functions defined on phase space. Below we shall see that the correspondence is consistent with the way canonical transformations act on functions.

In symbols, given $\tilde{H}(Q,P)$ and performing the transformation $(q,p) \to (Q,P)$, we obtain the transform of $\tilde{f}$ by the usual pull-back $f(q,p) = \tilde{f}(Q(q,p), P(q,p))$. Let $F(q,P)$ be the generating function of the transformation $(q,p) \to (Q,P)$.

Let $\tilde{\Phi}(Q)$ be a given function of the Q-coordinates and put $\Phi(q) = (T_F\,\tilde{\Phi})(q)$ according to the theory of chapter III. As done in section 1, we can define both $(\tilde{f}_{op}(Q,P)\,\tilde{\Phi})(Q)$ and $(f_{op}(q,p)\,\Phi)(q)$ and it is natural to expect that

$$(f_{op}(q,p)\,\Phi)(q) = (T_F(\tilde{f}_{op}(Q,P)\,\tilde{\Phi}))(q). \tag{4.1}$$

Although this seems to be rather hard to prove in general, we shall present another way of looking at it which makes it quite obvious. We shall also verify its validity in several cases.

As at the beginning of this chapter, let $\tilde{W}(Q,\bar{P},s)$ be the generating function of a monoparametric (one parameter) group of transformations. The function $\tilde{W}(Q,\bar{P},s)$ is to satisfy

$$\frac{\partial \tilde{W}}{\partial s} + \tilde{f}(Q, D_Q\,\tilde{W}) = 0, \qquad \tilde{W}(Q,\bar{P},0) = Q\,.\,\bar{P}.$$

We can now rewrite the right hand side of (4.1) as

$$(\partial T_F T_{\tilde{W}(s)} \tilde{\Phi})(q)/\partial s\,|_{s=0} = \partial T_{W(s)} \Phi(q)/\partial s\,|_{s=0}\,. \tag{4.2}$$

If we put $W(s) = \tilde{W}(s) \circ F$. Then (4.1) combined with (4.2) asserts that $W(q,\bar{p},s)$ must satisfy the equation

$$\frac{\partial W}{\partial s} + f(q, D_q W) = 0 \qquad W(q,\bar{p},0) = q \cdot \bar{p} \tag{4.3}$$

with $f(q,p) = \tilde{f}(Q(p,q), P(q,p))$

But we already know why this must be so! If we think of $\tilde{f}(Q,P)$ and $f(q,p)$ as Hamiltonians, (for they generate Hamiltonian flows according to (II-6.4)), the transformation generated by $F(q,P)$ is canonical, and (4.3) must hold!

Let us begin by verifying that everything works out as expected in a few cases. Let $F(q,P) = P \cdot \varphi(q)$ and firstly consider $\tilde{f} = \tilde{f}(Q)$. We have $\tilde{W}(s) = Q \cdot \bar{P} - \tilde{f}(Q)\, s$ and therefore $W(q,p,s) = \tilde{W} \circ F(q,\bar{p},s) = \varphi(q) \cdot \bar{p} - \tilde{f}(\varphi(q))\, s$ according to the composition rule (II-2.6). Therefore according to (4.1) - (4.3)

$$(f_{op}\, \Phi)(q) = \tilde{f}(\varphi(q))\, \tilde{\Phi}(\varphi(q))$$

as expected.

Consider again the case $F(q,P) = \varphi(q) \cdot P$ but now $\tilde{f} = \tilde{f}(P)$. This time $\tilde{W}(Q,\bar{P},s) = Q \cdot \bar{P} - \tilde{f}(P)\, s$ and $w(q,\bar{p},s) = \varphi(q) \cdot \tilde{f}(\bar{p})\, s$ which yields

$$(f_{op}\, \Phi)(q) = \int \tilde{f}(ik) \exp(-ik \cdot \varphi(q))\, \hat{\Phi}(k)\, \frac{dk}{(2\pi)^n}$$

Notice now $ik_j \exp(-ik \cdot \varphi) = -i\, J_{j\ell} \dfrac{\partial \exp(-ik \cdot \varphi)}{\partial q_\ell}$ with $J_{j\ell} = \dfrac{\partial \psi_\ell}{\partial Q_j}(\varphi(a))$ and $\psi(Q) = \varphi^{-1}(Q)$. With this

$$(f_{op}\, \Phi)(q) = \tilde{f}(-J\, \nabla_q)\, \Phi(\varphi(q))$$

which is what we would expect.

Consider now $F(q,P) = q \cdot P - g(q)$ and we examine again the case $\tilde{f} = \tilde{f}(Q)$. In this case $\tilde{W}(Q,\bar{P},s) = Q \cdot \bar{P} - \tilde{f}(Q)s$ and $W(q,\bar{p},s) = q \cdot \bar{p} - g(q) - \tilde{f}(q)s$. Therefore

$$(\tilde{f}_{op}\, \tilde{\Phi})(Q) = \tilde{f}(q)\, \Phi(q)$$

which is alright since the effect of $F(q,P) = q \cdot P - g(q)$ is to get $Q = q$ and $P = p - \nabla g(a)$

Consider now for the same $F(q,p)$ but with $\tilde{f} = \tilde{f}(p)$. This time we obtain $W(q,\bar{p},s) = q \cdot \bar{p} - g(q) - \tilde{f}(p)\, s$ and

$$(f_{op}\,\Phi)(q) = \int f(ik) \exp(-ik \cdot q + g(a))\, \hat{\tilde{\Phi}}(k) \frac{dk}{(2\pi)^n}$$

$$= \int \exp(g(a))\, (f(-D) \exp(-ik \cdot a))\, \hat{\tilde{\Phi}}(k) \frac{dk}{(2\pi)^n}$$

$$= f(-D + \nabla g(a))\, \Phi(q)$$

where $\Phi(q) = \exp(g(q))\tilde{\Phi}(q)$ and we used $\exp(g(q))(-D)\exp(-g(q)) = (-D + \nabla g)$ as operator on fucntions $\Phi(q)$.

Let us consider now $F(q,P) = q \cdot P - h(P)$ and let us again examine the effect of the transformation $Q = q - \nabla h(P)$, $P = p$ on $\tilde{f} = \tilde{f}(Q)$ to begin with. In this case, in order to compute $W(q,p,s)$ we would have to solve

$$\begin{cases} q_1 = q - \nabla_p h(p_1) \\ p_1 = \bar{p} - \nabla_q \tilde{f}(q_1)\, s \end{cases} \quad \text{or} \quad \begin{cases} q_1 + \nabla h(p_1) = q \\ p_1 + \nabla\tilde{f}(q_1)\, s = \bar{p}\,. \end{cases}$$

Instead of struggling with this we are going to verify (4.1) directly. The transform of $\tilde{f}(Q)$ is $\tilde{f}(q - \nabla h(p)) = f(q,p)$. Consider now, with $X(D) = \nabla h(-D)$

$$f(q - X(D))\, (T_F\, \tilde{\Phi})(q) = \int\int \hat{\tilde{f}}(l) \exp(-il \cdot (q - X(D)))$$

$$\exp(-ik \cdot q)\, (T_F\, \tilde{\Phi})^{\wedge}(k)\, dl \frac{dk}{(2\pi)^{2n}}$$

we know that $(T_F\, \Phi)^{\wedge}(k) = \exp h(ik)\, \hat{\tilde{\Phi}}(k)$ and from (3.4) we have

$$\exp - il \cdot (q - X(D))\, e_k(q)$$

$$= \exp\{h(ik + il) - h(ik)\} \exp q \cdot (-il - ik)$$

and therefore

$$f(q - X(D))\,(T_F\,\tilde{\Phi})(q)$$

$$= \int\int \hat{f}(l) \exp(h(ik + il)) \exp(-il \,.\, q) \exp(-iq \,.\, k)\, \Phi(k)\, dk \frac{dl}{(2\pi)^n}$$

ook at $T_F(\tilde{f}_{op}\,\tilde{\Phi})(q)$. Recall that

$$\int \exp(-ik\, Q)\, \tilde{f}(Q)\, \tilde{\Phi}(Q) = \int \hat{f}(k - \bar{k})\, \hat{\tilde{\Phi}}(\bar{k}) \frac{d\bar{k}}{(2\pi)^n}$$

herefore

$$(T_F\,\tilde{f}\,\tilde{\Phi})(q) = \int \exp(-ik \,.\, q + h(ik)) \left(\int \hat{f}(k - \bar{k})\, \hat{\tilde{\Phi}}(\bar{k}) \frac{d\bar{k}}{(2\pi)^n}\right) \frac{dk}{(2\pi)^n}$$

hich is $f(q - X(D))(T_F\,\tilde{\Phi})(q)$ after a trivial change of variables.

The case $\tilde{f}(D)$ is easy since $f(q,p) = \tilde{f}(P)$. This time $W(q,\bar{p},s) = q \,.\, p - h(P) - \tilde{f}(p)s$ and e obtain

$$f_{op}\,(T_F\,\Phi)(q) = (T_F\,\tilde{f}_{op}\,\tilde{\Phi})(q)$$

$$= \int f(ik) \exp(-ik \,.\, q + h(ik))\, \hat{\Phi}(k) \frac{dk}{(2\pi)^n}$$

Also for later use, and here we depart from the simple cases, consider $F(q,P) = q \circ \varphi(P)$ nd $\tilde{f} = \tilde{f}(P)$. Now, repeating the procedures outlined above we obtain $W(q,\bar{p},s) =$ $\varphi(\bar{p}) \,.\, q - \tilde{f}(p)\, s$ which yields

$$(f_{op}\,T_F\,\tilde{\Phi}) \equiv (f_{op}\,\Phi)(q) = \int \tilde{f}(ik) \exp(-q \,.\, \varphi(ik))\, \hat{\tilde{\Phi}}(k) \frac{dk}{(2\pi)^n}$$

$$= f(\psi(-D))\,(T_F\,\tilde{\Phi})\,(q).$$

We shall state without proof the results of applying any of the two sides of (4.1) to the perators $\hat{f}$ associated to functions $f(q,p)$. Recall that to $\tilde{f}(Q,P)$ we associated the operator

$$(\hat{\tilde{f}}_{op}(Q,P)\,\tilde{\Phi})(Q) = \int \tilde{f}(Q,ik) \exp(-ik \,.\, Q)\, \hat{\tilde{\Phi}}(k) \frac{dk}{(2\pi)^n}.$$

Let now $\tilde{f}(Q,P) = Q^m P^n$ throughout the remainder of the section. Let $\tilde{\Phi}(Q)$ be giv
and begin with $F(q,P) = \varphi(q) \cdot P$. In this case

$$T_F(\hat{\tilde{f}}_{op}(Q,P)\,\tilde{\Phi})(q) = (\hat{f}\, T_F\, \Phi)(q) = (\varphi(q))^m\, (J \cdot \nabla)^n\, \tilde{\Phi}(\varphi(q))$$

with $J_{j\ell} = (\partial\psi_\ell/\partial Q)\,(\varphi(q))$ and $\psi = \varphi^{-1}$.

When $F(q,P) = q \cdot P + g(q)$ we obtain

$$T_F(\hat{\tilde{f}}_{op}\,\tilde{\Phi})(q) = (\hat{f}_{op}\, T_F\, \Phi)(q) = q^m\, (-D - \nabla g)^n\, \Phi(q)$$

with $\Phi(q) = \exp(-g(q))\,\tilde{\Phi}(q)$.

Consider now the case $F(q,P) = q \cdot P - h(P)$. Now the result is

$$T_F\,(\hat{\tilde{f}}_{op}\,\tilde{\Phi})(q) = \hat{f}_{op}\, T_F\, \Phi(q) = (q - \nabla h(-D))^m\, (-D)^n\, \Phi(q)$$

with $\Phi(q)$ such that $\hat{\Phi}(k) = \exp h(ik)\, \hat{\tilde{\Phi}}(k)$.

And to finish, we have $F(q,P) = \varphi(P)$. We now obtain

$$T_F\, \tilde{f}_{op}\, \tilde{\Phi}(q) = \hat{f}_{op}(T_F\, \Phi)(q) = (q^T J(\psi)(-D)))^m\, (\psi(-D))^n\, (T_F\, \tilde{\Phi})(q).$$

In other words the ordered product correspondence rule $\tilde{f}(Q,P) \rightarrow \hat{\tilde{f}}_{op}$ given by (2.1) such that our expectations (given by 4.1) are fulfilled when F is one of the types consider in the list.

₹eferences

√-1] Feinsilver, P. J. : "Special Functions, Probability Semigroups and Hamiltonian Flows". Springer-Verlag. L.N.M. 696, Berlin, 1978.

√-2] Feinsilver, P. J. : "An Introduction to Operator Calculus and Orthogonal Theory". Book to appear.

ther correspondence rules can be seen in

V-3] Weyl, H. : "The Theory of Groups and Quantum Mechanics". Methuen, London, 1931.

V-4] Moyal, J. E. : "Quantum Mechanics as a Statistical Theory". Proc. Cambridge Philosophical Soc. **45**, (1949) 99-124.

ome papers and books on operator calculus are

V-5] Mc. Cay, N. H. : "On Commutator Rules in the Algebra of Quantum Mechanics". P.N.A.S., U.S. **5** (1929), 200-202.

V-6] Wilcox, R. M. : "Exponential Operators and Parameter Differentiation". Jour. Math. Phys. **8**, N} 4 (1967), 962-982.

V-7] Maslov, V. P. : "Operational Methods". MIR Publishers, Moscow, 1976.

V-8] Jorgensen, P. W. T. and Moore, R. T. : "Operator Commutation Relations". D. Reidel Pub. Co. Dordrecht, 1984.

or an update on Lie-Trotter product formulas take a look at references [V.9] and [V.10].

Chapter V
Evolution Semigroups

V.1 Evolution Semigroups

Evolution semigroups or propagators appear when one studies the dependence of the solutio
u(q,t) of an equation like

$$\frac{\partial u}{\partial t}(q,t) = (G\,u)(q,t), \qquad t > 0, \quad q \in \mathbb{R}^n$$

$$u(q,0) = \Phi(q) \tag{1.1}$$

on the initial datum $\Phi(q)$. The evolution semigroup will be a family $\{P_t\}_{t\geq 0}$ mapping th
initial data $\Phi(q)$ onto the present data $u(q,t)$.

We shall assume that when the initial $u(q,0)$ is in the class of test functions, the solutio
$u(q,t)$ of (1.1) exists and, moreover it is also a test function. We shall also assume that
family of operators $\{P_t\}_{t\geq 0}$ exists such that

$$u(q,t) = (P_t\,\Phi)(q) \tag{1.2}$$

for test functions $\Phi(q)$, and that it satisfies

(i) $P_t\Phi \to \Phi$ as $t \downarrow 0$.

(ii) $P_{t+s} = P_t\,P_s$ all $t,s \geq 0$.

(iii) $\dfrac{\partial u}{\partial t} = G\,P_t\Phi = G\,u.$

Property (iii) is rephrased by saying that G is the infinitesimal generator of P_t.

Whenever f(q) is only a locally integrable function, we shall give meaning to G f an
$P_t\,f$ by means of the duality relationship introduced in chapter III.

We shall use the same symbol for G *as an operator on test functions or on their dual.*

In this chapter we are going to examine the interplay between the theory of represen
tations of canonical transformations and the operational calculus that can be used to solv
equations like (1.1).

We are going to consider infinitesimal generators $G(q,\Delta) = K(\Delta) + V(q)$ which is th al of $K(-\Delta)+V(q)$ when it acts on S, the class of test functions described in chapter III.

The expression

$$(P_t f)(q) = \int \rho_t(q,q') f'(q') dq' \tag{1.3}$$

hall have only symbolic meaning, because the density $\rho_t(q,q')$ may not be a function.

Our basic assumption is

ssumption 1.4 *There exists a positive function* $\Omega_0(q)$ *such that* $P_t \Omega_0(q) = \Omega_0(q)$, *and* *erefore* $G\Omega_0(q) = 0$.

omment When no such $\Omega_0(q)$ exists, but there exists a positive $\Omega_0(q)$ such that $\Omega_0(q) = \lambda\ \Omega_0(q)$, for some λ, we shall replace P_t by $\exp - \lambda t\ P_t = \bar{P}_t$, then $\Omega_0(q) = \Omega_0(q)$. Now $\tilde{G} = G - \lambda$.

efinition 1.5 A function $\Omega_0(q)$ such that (1.4) holds will be called a *vacuum* for P_t.

Consider now the canonical transformation generates by $F(q,P) = q \cdot P - \ln \Omega_0(q)$ under hich $Q = q, P = p + \nabla \ln \Omega_0$.

Then the semigroup $\{P_t\}$ acting on functions $f(q)$ is transformed into a semigroup $\tilde{P}_t$ cting on functions $\tilde{f}(q) = (T_F^{-1} f)(q)$ given by

$$\tilde{P}_t \tilde{f}(q) = (T^{-1} P_t \tilde{f})(Q) = \int \Omega_0(Q)^{-1} \rho_t(Q,Q') \Omega_0(Q') \tilde{f}(Q') d\tilde{Q}'$$

$$= \int \tilde{\rho}_t (Q,Q') \tilde{f}(Q') dQ' .$$

And we have

emma 1.6 *The family of operators* $\{\tilde{P}_t\}$ *defined by*

$$(\tilde{P}_t \tilde{f})(Q) = \frac{1}{\Omega_0(Q)} (P_t \Omega_0 \tilde{f})(Q)$$

a semigroup such that $\tilde{\Omega}_0(q) \equiv 1$ *is a vacuum for* $\tilde{P}_t$ *if and only if* $\Omega_0(q)$ *is a vacuum for* P_t.

roof For the proof of $\tilde{P}_{t+s} = \tilde{P}_t \tilde{P}_s$ notice that the right cancellatins occur.

To see that $\tilde{\Omega}_0(q) \equiv 1$ is a vacuum just take $\tilde{f} = 1$.

In the physical literature this transformation of semigroups is known as gau[illegible] transformation. Below we shall relate it to the Cameron–Martin–Girsanov transformation [illegible] probabiity theory.

We leave as an exercise the reader to verify that if we denote the infinitesimal generat[illegible] $\tilde{P}_t$ by $\tilde{G}$, it is related to the generator G of P_t by

$$\tilde{G}\,\tilde{f}(Q) = \frac{1}{\Omega_0(Q)}\, G(\Omega_0 f)(Q)$$

and we have

Lemma 1.3 *The solutions to*

$$\frac{\partial \tilde{u}}{\partial t} = \tilde{G}\,\tilde{u} \qquad\qquad \frac{\partial u}{\partial t} = G\,u$$

are related by

$$(T_F^{-1}\, u(\cdot\,,t))(Q) = \tilde{u}(Q,t).$$

Comment This is a restatement of lemma (1.2) and just for emphasis on meaning.

V.2 Semigroups and Hamiltonian Systems: The Operational Calculus Connection

Here we carry out an extension of the results of Feinsilver's [IV–1]. We provide a differe[illegible] way of looking at things and a framework in which the results can be extended.

The fundamental result of this section is contained in

Theorem 2.1 *Let* $C_1^+(t)$ *be the operators satisfying*

$$C_1^+(t) = [\,G, C_1^+(t)\,] \qquad (C_1^+(0)\, f)(q) = q\, f(q)$$

for $i = 1,2,\ldots,n$. *Here* G *is obtainable from* $K(p) + U(q) = H(q,p)$ *by* $G = K(\Delta) + U(q$[illegible] *Then*

$$\tilde{P}_t\, f)(Q) = \Omega_0(Q)^{-1}\, f(C^+(t))\, \Omega_0(q) \tag{2.[illegible]}$$

for f *having appropriate power series expansion*

roof Note that the soluton to $C_1^+(t) = [G, C_1^+]$ is given by

$$C_1^+(t) = \exp t\, G\, q_i \exp - t\, G \equiv \exp t\, [G, \cdot\,]\, q_i \tag{2.3}$$

d therefore

$$\tilde{P}_t \tilde{f}(Q) = \Omega_0(Q)^{-1} (\exp(tG) f\, \Omega_0)(Q)$$

$$= \Omega_0(Q)^{-1} (\exp t\, G\, f \exp - t\, G)\, \Omega_0(Q)$$

$$= \Omega_0^+(Q)\, f(C^+(t))\, \Omega_0(Q)$$

here for the second identity we invoked assumption (1.4) and for the third we used (2.3) mbined with the analyticity assumption on f.

omments

(a) The operator $[G, \cdot\,]$ is called the Liouville operator in the physical literature and, Ad G in the mathematical literature.

(b) It usually happens that $\exp -tG$ is not well defined. No problem, for

$$(\exp t\, G\, f)(q) = (\exp t\, [G, \cdot\,]\, f) \exp t\, G$$

may still make sense. Here $f(q)$ is to be considered a multiplicative operator. See [IV-8] for an indepth study of such problems.

(c) In the physical literature the operators $C_1^+(t)$ are called creation operators and the corresponding $C_i(t) = \exp tG \left(\frac{\partial}{\partial q_i} \right) \exp -tG$ are called annihilation operators. We shall see why when we study moment systems

The problem now is how to express $C_1^+(t)$ in terms of q, ∇_q and objects associated with e Hamiltonian system.

As a heuristic guide, consider the case treated by Finsilver in [IV.1].

roposition 2.4 *Let* $G = H(\nabla)$ *and assume that* $H(0) = 0$. *Then* $C^+(q) = q + t\, v(\nabla_q)$ *with* $(p) = \nabla_p H(p)$ *and* (2.2) *is trivially verifiable with* $\Omega_0(q) = 1$.

Proof We proved that $[H(\nabla_q), q_i] = v_i(\nabla_q)$ in chapter IV. Therefore, since $v(\nabla_q)$ commutes with $H(\nabla_q)$ the successive applications of $[H(\nabla), \cdot]$ to q vanish after the firs and we obtain

$$C_1^+ = \exp t [H(\nabla),] q = q_i + t\, v_i(q).$$

To verify that (2.2) holds, note that

$$f(C^+(t))\, 1(q) = \int \hat{f}(k) [\exp (ik \,.\, C^+(t)\, 1] (q) \frac{dk}{(2\pi)^n}$$

$$= \int \exp (ik \,.\, q + t\, H(ik))\, \hat{f}(k) \frac{dk}{(2\pi)}$$

where we used the fact that $1 = \exp i\, 0 \,.\, q$ and formula (3.4) was properly adapted.

Comment Consider the Hamilton–Jacobi function $S(q,P,t) = q \,.\, P - t\, H(p)$ applied to Φ by means of T. We know that $T_{S(t)} \Phi(q)$ satisfies $\frac{\partial}{\partial t} T_{S(t)} \Phi = H(-\nabla)\, T_{S(t)} \Phi\, S(t)$. Therefore if

$$< P_t\, f, \Phi > \equiv < T_{S(t)}\, f, \Phi > = < f, T_{S(t)}\, \Phi >$$

we have that

$$\frac{\partial P_t\, f}{\partial t} = H(\nabla)\, P_t\, f.$$

In this case (2.2) is readily verifiable, for

$$< f, T_{S(t)}\, \Phi > = < 1, f_{op}\, T_{S(t)}\, \Phi > = < T_{S(t)}^{-1} T_S^{-1}\, f_{op}\, T_S\, \Phi >$$

$$= < P_t\, 1, T_{S(t)}^{-1} f_{op}\, T_{S(t)}\, \Phi > = < 1, f_{op}\, (C^+(t)) \Phi >.$$

We leave for the reader to verify the last step using the results of chapter IV. Note tha we act on test functions $C^+(t) = q + t\, V(-\nabla)$ and only when we dualize we obtain the resul in proposition (2.4).

The annihilation operators $C_i(t)$ satisfy

Lemma 2.5 *Define* $C_i(t) = \exp t [G, \cdot]\, \nabla_q \exp t\, G$ *then*

$$\dot{C}_i(t) = [\,G,C_i\,].$$

roof Simple.

Two equally simple to understand results, basic to the establishment of the analogue of roposition 2.4 are the following

emma 2.6 *The components* $C_1^+(t)$, $C_j(t)$, $1 \leq i,\ j \leq n$, *of* $C^+(t)$, $C(t)$ *satisfy*

$$[\,C_1^+(t), C_j(t)\,] = \delta_{ij} \qquad 1 \leq j,\ i \leq n$$

or all t.

roof Since $[\,C_j(0), C_1^+(0)\,] = \left[\dfrac{\partial}{\partial q_j}, q_i\right] = \delta_{ij}$, after pre- and postmultiplication by xp t G and exp – t G respectively we obtain the desired conclusion.

emma 2.7 *The equations* $\dot{C}^+(t) = [\,G, C^+(t)\,]$ *and* $\dot{C}(t) = [\,G, C(t)\,]$ *can be written as*

$$\dot{C}^+(t) = [\,G, C^+(t)\,] = (\nabla_p K)\,(C(t))$$
$$C^+(t) = [\,G, C(t)\,] = -\,(\nabla_q U)\,(C^+(t)). \tag{2.8}$$

roof Using $G = (\exp tG)G(\exp - tG)$ we can rewrite G as $G = K\,(C(t)) + U(C^+(t))$. Jsing lemma 2.8 and the commutation relation derived from the GLL we obtain

$$[\,G, C^+(t)\,] = [\,K(C(t)), C^+(t)\,] = (\nabla_p K)\,(C(t))$$

$$[\,G, C^+(t) = = [\,U(C^+(t)), C(t)\,] = -\,(\nabla_q U)\,(C^+(t))$$

which is the result we want.

Let us now introduce some notations. Let us write the solution to the Hamiltonian system s a flow

$$\Phi(t) \equiv (\Phi^{(1)}(t), \Phi^{(2)})\,;\ \mathbb{R}^{2n} \to \mathbb{R}^{2n}$$

which is family of diffeomorphisms satisfying $\Phi(t+s, q,p) = \Phi(t, \Phi(s,q,p))$. Let us denote the irst n components of Φ by $\Phi^{(1)}$, and the second n components by $\Phi^{(2)}$.

Recall that to a function $X(q,p)$ we associated an operator $\hat{X}_{op}$ on test functions $\psi(q)$ y

$$(\hat{X}_{op}\psi)(q) = \int X(q,ik)\exp(-ik\,.\,q)\hat{\psi}(k)\frac{dk}{(2\pi)^n}.$$

This action can be transported to the dual space by

$$(\hat{X}_{op}\, f)(q) = \int \exp(ik\,.\,q)\,(\hat{X}\, f)(k)\frac{dk}{(2\pi)^n} \tag{2.4}$$

where

$$(\hat{X}\, f)^{\wedge}(k) = \int f(q)\, X(q,ik)\exp(-ik\,.\,q)\, dq. \tag{2.10}$$

By means of this correspondence we obtained, for example, that for $X = X(q)$ (or $X = X(p)$) we had $(\hat{X}_{op}\, f)(q) = X(q)\, f(q)$ (or $(\hat{X}\, f)(q) = X(D)\, f(q)$).

Now comes the description of $C^+(t), C(t)$ in terms of the flow.

Theorem 2.11 *Define operators* $\hat{\Phi}_i^{(1)}(t)$, $\hat{\Phi}_j^{(2)}(t)$ *by means of* (2.9)-(2.10), *with the* Φ_j *being the components of the flow* $\Phi(t)$ *on* $\mathbb{R}^{2n}$ *that solves the Hamilton equations.* *Assume that*

$$[(\nabla K(\Phi^{(2)}(t))]_{op} = \nabla K(\hat{\Phi}^{(2)}_{op}(t)) \quad (\nabla U)(\Phi^{(1)}_{op}(t)) = [\nabla U\hat{\Phi}^{(1)}_{op}(t))]. \tag{2.12}$$

Then

$$C_1^+(t) = \Phi^{(1)}_{op}(t) \quad and \quad C_j(t) = \Phi^{(2)}_{op}(t) \tag{2.13}$$

Proof Notice to begin with that both sides of (2.13) coincide at $t = 0$. We shall now verify that the time derivatives of $\hat{\Phi}^{(1)}_{op}(t)$ and $\Phi^{(2)}_{op}(t)$ satisfy (2.8). Start with (2.10) applied to $\hat{\Phi}^{(1)}(t)$.

$$(\Phi_{op}\hat{(t+\varepsilon)}\, f)(k) = \int f(q)\,\hat{\Phi}^{(1)}_{op}(t+\varepsilon,ik)\exp(-ik\,.\,q)\, dq$$

$$+\,\varepsilon\int f(q)\,(\nabla K)\,(\Phi^{(2)}_{op}(t,q,ik))\exp(-ik\,.\,q)\, dq + O(\varepsilon)$$

$$= (\Phi^{(1)}_{op}\, f)(\hat{k}) + \varepsilon[(\nabla_p K(\Phi^{(2)}_{op})f]^{\wedge}(k) + O(\varepsilon)$$

$$= (\hat{\Phi}^{(1)}_{op}\, f)^{\wedge}(k) + \varepsilon\,[\nabla_p K(\Phi^{(2)}_{op})\, f]^{\wedge}(k) + O(\varepsilon)$$

where assumption (2.12) was brought in for the last step. Now

$$\left(\frac{\partial\hat{\Phi}^{(1)}(t)f}{\partial t}\right)^{\wedge}(k) = [\,(\nabla_p K)\,(\Phi^{(2)}_{op})\,f\,]^{\wedge}(k)$$

or $\dfrac{\partial\hat{\Phi}^{(1)}_{op}}{\partial t} = (\nabla_p K)(\Phi^{(2)}_{op})$ as operators. The same reasoning applied to $\hat{\Phi}^{(2)}(t)$ yields

$$\frac{\partial\hat{\Phi}^{(2)}_{op}}{\partial t} = -\,(\nabla U)(\hat{\Phi}^{(1)}(H)).$$

Comment Even though assumption (2.12) is very strong (you may say: why don't we assume the theorem to begin with?) it can still be verified in a variety of cases. The most important one of course corresponds to quadratic Hamiltonians.

To close this section let us take another look at (2.2) from another point of view. It follows from (2.8) that

$$\frac{df}{dt}(C^{+}(t)) = [\,G,f(C^{+}(t))\,]$$

and since $G\,\Omega_0(q) = 0$ we obtain $\dfrac{\partial f(C^{+}(t)\Omega_0)(q)}{\partial t} = G(f(C^{+}(H))\,\Omega_0)(q)$. Now, premultiplying both sides by $\Omega_0(q)^{-1}$ and using $\tilde{G} = \Omega_0^{-1}\,G\,\Omega$ and $\tilde{P}_t = \Omega_0^{-1}\,P_t\,G$ we reobtain $\dfrac{\partial\tilde{P}_t f(q)}{\partial t} = \tilde{G}\,\tilde{P}_t\,f(q)$.

V.3 Generalized Markov Processes and Martingales

In this section we introduce some basic terminology in the theory of Markov processes, and we obtain an extended version of the Feynman-Kac formula.

The basic ingredients for the specification of a Markov process are a sample space $(\Omega, \mathcal{F})$, a state space $(E,\mathcal{E})$, a family of coordiante mappings $\{X(t) : \Omega \to E \mid t \in (0,\infty)\}$ and a family $\{P^{x} : x \in E\}$ of probability measures on $(\Omega, \mathcal{F})$.

Here we shall take Ω as being the collection of all right continuous functions $w(t) : [0,\infty) \to E$ which have left limits. $\mathcal{F}$ denotes the σ-algebra of subsets generated by the coordinate maps $X(t)(w) = w(t)$ and we take $E = \mathbb{R}^n$ and $\mathcal{E} = \mathcal{B}(\mathbb{R})$, the Borel subsets of $\mathbb{R}^n$. Thus Ω contains all the "rando motion" of the system and $\mathcal{F}$ all the "questions we can ask about the system".

The other object which is handy to have around is the family $\mathcal{F}_t = \sigma\{X(s) : s \le t\}$. Each of these sub-algebras contains the information about the system up to time t.

To obtain generalized Markov processes we are going to assume that we have a semigroup of operators $\{P_t : t > 0\}$ as specified in the previous section, having the *Marko* *property*

$$P_t\, 1(q) \equiv 1$$

but not necessarily positive, i.e. $P_t\, f \not\geq 0$ for arbitrary $f \geq 0$.

For any initial probability measure on $\mathbb{R}^n$ and any finite collection $J \neq \{0 \leq t_1 < \ldots < t_m$ we *define* for $B_1,\ldots,B_m$ in $B(\mathbb{R})$

$$P_J^{\mu}(X(t_1) \in B_1,\ldots,X(t_n) \in B_n)$$

$$= \int_{\mathbb{R}^n} \mu(dq_0) \int_{B_1} P_{t_1}(q_0\, dq_q) \int_{B_2} P_{t_2 - t_1}(q_1, dq_2) \ldots \int_{B_m} P_{t_m - t_{m-1}}(q_{n-1}, dq_m)$$

or for any bounded $f(q_1,\ldots,q_m) : \mathbb{R}^n \times \overset{m}{\ldots} \times \mathbb{R}^n \to \mathbb{R}$,

$$E_J^{\mu} f(X_{t_1}, \ldots, X_{t_n})$$

$$= \int \mu\, (dq_0) \int \ldots \int P_{t_{12}}(q, dq_1)\, P_{t_2 - t_1}(q_1, dq_2) \ldots P(q_{m-1}, dq_m)\, f(q_1,\ldots,q_m). \tag{3.2}$$

It is easy to verify that the family of measures so defined is consistent in the followin sense. A function $f(q_1,\ldots,q_m)$ may be thought of as a function $f(q_1,\ldots,q_k)$, with $k > m$, bu independent of $q_{m+1},\ldots,q_k$. Let $J_1 = \{t_1 < t_2 < \ldots < t_m\}$ and $J = \{t_1 < t_2 < \ldots < t_m < \ldots < t_k\}$ Then

$$E_{J_1}^{\mu} f(X(t_1),\ldots,X(t_m)) = E_{J_2}^{\mu}(f\,(X(t_1),\ldots,X(t_m)).$$

All that it takes to verify such identities is the Markov condition $P_t\, 1 = 1$ imposed on th semigroups. When the semigroups P_t are positive, the family E_J^{μ} can be uniquely extende to a σ-additive measure P^{μ} on the measure space (Ω, F).

When the P_t are not positive operators as will be our case, we can only obtain a finitel additive set function P^{μ} on (Ω, F). A lot can be done with this anyway. For any randon variable, i.e., measurable function $H : \Omega \to \mathbb{R}$, its expected value (when it exists) with respec to P^{μ} is denoted by

$$E^{\mu}(H) = \int H(w)\, dP^{\mu}(w)$$

which is sometimes called path integral or functional integral. For the statement of th

Markov property we need to introduce the concept of conditional expectations.

Definition 3.3 Let G denote a sub σ-algebra of F and let H be a random variable, integrble with respect to a measure P^{μ}. The conditional expectation of H with respect to G is the G-measurable random variable $E^{\mu}[H \mid G]$ satisfying

$$E^{\mu}[F H] = E^{\mu}[F E^{\mu}[H \mid G]]$$

for every bounded G-measurable function F.

When P^{μ} is not a measure but only a finitely additive function we still have

Definition 3.3-a The conditional expectation of H with respect to G (relative to P^{μ}) is the random variable $E_c^{\mu}[H \mid G]$, measurable with respect to G, such that

$$E^{\mu}[F G] = E^{\mu}[F E_c^{\mu}[H \setminus G]]$$

for every cylindrical function F.

Comment Cylindrical functions are those that belong to a cylindrical σ-algebra, i.e., a σ-algebra generated by finitely many $X(t)$'s.

So, assume from now on that we have constructed, starting from a generalized Markov semigorup P_t, and a measure μ on $\mathbb{R}^n$, a (possibly only finitely additive) measure P^{μ} on (Ω, F). We have rigged up this set in order to have

Proposition 3.4 *The process* $X(t)$ *is a generalized Markov process, i.e., for every* $s, t \geq 0$ *and bounded* $f(q)$ *we have*

$$E_c^{\mu}[f(X(t+s)) \mid F_t] = (P_s f)(X(t)) = \int P_s(X(t), dy) f(y).$$

Proof Just apply the definitions of conditional expectation and (3.1) to obtain the desired result.

Comments

a) Whenever the P^{μ} are countably additive measures, then

$$E_c^{\mu}[f(X(t+s) \mid F_t] \text{ becomes } E^{\mu}[f(X(t+s)) \mid F_t].$$

b) We would be quite inconsistent, wouldn't we, if we dragged out the usual a.s. - qualifiers all over the place!

Definition 3.5 A family $\{Z_t : t \geq 0\}$ of measurable and integrable random variables Z is called a Martingale (or an extended Martingale) if

(i) Z_t is F_t-measurable for each t.

(ii) $E^{\mu}[Z_{t+s} \mid F_t] = Z_t$ or $E_c^{\mu}[Z_{t+s} \mid F_t] = Z_t$

depending on whether P^{μ} is a measure of a finitely-additive set function.

We list three generic types of Martingales.

For f in the domain of G we have that

$$Z(t) = \frac{f(X(t))}{f(X(0))} \exp - \left\{ \int_0^t \frac{Gf(X(s))}{f(X(s))} ds \right\} \tag{3.9}$$

whenever f never vanishes, also

$$M(t) = f(X(t)) - f(X(0)) - \int_0^t (G\, f)\, (X(s))\, ds \tag{3.10}$$

is a Martingale. For f(t,q) being differentiable with respect to t for each q and in the domain of G for each t.

$$X(t) = f(t,X(t)) - f(0,X(0)) - \int_0^t \left(\frac{\partial}{\partial s} + G \right) f(s,X(s))\, ds$$

is a Martingale. In chapter VI we present some examples.

V.4 The Generalized Feynman-Kac Formula

We have developed enough notation and terminology to take up again an unconcluded theme from section 4, chapter III.

Recall that we obtained S(q,P,t) as the limit of a sequence $S^{(N)}(q,P,t)$ which was itself given as the composition

$$S^{(N)}(q,P,t) = S_N^{N-1} \circ S_N^{N-2} \circ \ldots \circ S_N^{0}(q,P,t)$$

with each of the factors being of the type

$$S_N^{(k)}(q^{k-1},p^k) = (F_k \circ F_v)(q^{k-1}, p^k) = q^{k-1} \,.\, p^k - \frac{t}{N}(K(p^{(k)}) + V(q^{(k-1)})).$$

To obtain the variant of the Feynman–Kac formula we denote by $X(t)$ the generalized process with homogeneous, independent increments on $\mathbb{R}^n$, with infinitesimal generator $G = H(\nabla)$ and semigroup $Q_t f(q)$ such that

$$< Q_t{}^f \Phi > = < f, T_{F_k(t)} \Phi > \quad \text{with} \quad F_k(q,P,t) = q \,.\, P - K(P)\, t.$$

The obvious direct representation of Q_t is given by

$$(Q_t f)(q) = (T_k f)(q) = \int \exp \{ik \,.\, q + t\, K(ik)\} \hat{f}\,(k) \frac{dk}{(2\pi)^n} \tag{4.1}$$

with the $f's$ in the dual of the test functions being Fourier transformed as $\hat{f}(k) = \int (\exp\ -ik \,.\, q)\, f(q)\, dq$. From (4.1) we see that the semigroup has a density whenever

$$\rho_t(q,q') = \int \exp \{ik \,.\, (q - q') + t\, K(ik)\} \frac{dk}{(2\pi)^n} \tag{4.2}$$

is a well defined function. Recall that we are assuming $K(0) = 0$, so that $Q_t\ 1 = 1$. Now by dualization

$$< f, T_{S_N^{(N)}} \Phi > \equiv < T_{S_N^{(N)}} f, \Phi > = < \prod_{j=1}^{n} T_{F_k} T_{F_v} f, \Phi >.$$

If we denote by $\{P^q : q \in \mathbb{R}^n\}$ the measure induced on (Ω, G) by the semigroup Q_t we have

$$(T_F\, T_v\, f)(q) = E^q \{\exp [\frac{t}{N} V(X\left(\frac{t}{N}\right))]\, f(X\left(\frac{t}{N}\right)\}$$

and if we iterate, the result of applying $T_{S_N^{(N)}}$ to f is given by

$$(T_{S_N^{(N)}(t)}\, f)(q) = E^q \left\{ \exp\left[\sum_1^n \frac{t}{N} V\left(X\left(\frac{tk}{N}\right)\right) \right] f(X(t)) \right\}$$

Here is where we make use of our choice of path space: as N tends to infinity the exponent under the integral converges to the limit $\exp \left\{ \int_0^t V(X(s))\, ds \right\}$ and we obtain that

$$\lim \left(T_{S_N^{(N)}(t)} f\right)(q) = E^q \left\{ \exp \int_0^t V(X(s))ds f(X(t)) \right\}$$

which is the famous Feynman-Kac representation formula for the solutions to the equation

$$\frac{\partial u}{\partial t} = Gu = (K(\nabla) + V(q))u \quad u(q,0) = f(q).$$

References

Apart from the material in [III.8] two sources with complementary material are

[V-1] Goldstein, J. : "Semigroups of Linear Operators and Applications". Oxford Univ. Press, Oxford, 1985..

[V-2] Dollard, J. D. and Friedman, C. N. : "Product Integration". Encyclopedia of Mathematics. Vol. 10. Addison-Wesley, Reading, 1079.

Connections between semigroups and probability and related topics are contained in

[V-3] Ethier, S. N. and Kurtz, T. G. : "Markov Processes". John Wiley & Sons, New York, 1985.

[V-4] Metivier, M. : "Limites Projectives de Mesures, Martingales, Appications". Ann. Math. Pura. Appl. IV. Ser. **63** (1963) 225-352.

Generalized Processes and some nonstandard material can be seen in

[V-5] Krylov, V. In. : "Some Properties of the Distribution corresponding to $\frac{\partial u}{\partial t} = (-1)^{1+1} \frac{\partial^{2q} u}{\partial x^{2q}}$". Sov. Math. Doklady **1**, (1970) 160-773.

[V-6] Hochberg, K. J. : "A Signed Measure on Path Spaces related to Wiener Measure". The Annals of Probability, **6** No. 3 (1978) 433-458.

[V-7] Nishioka, K. : "Stochastic Calculus for a Class of Evolution Equations". Jap. Jour. Math, **11** No. 1 (1985) 59-102.

[V-8] Berger, M. A. and Sloan, A. D. : "A Method of Generalized Characteristics". Mem. Amer. Math. Soc **266** (1982).

And on the Trotter-Lie formula we list

[V-9] Pazy, A. : "A Trotter-Lie Formula for Compactly Generated Semigroups of Non-linear Operators". J. Funct. Analysis 13 (1976) 353-361.

[V-10] Goldstein, J. A. : "Remarks on the Feynman-Kac Formula in Partial Diff. Equations and Dynamical Systems". Ed. Fizgibbon, W. E. Pitman Adv. Pub. Program., Boston, 1984.

Chapter VI
Applications

The different sections of this chapter contain examples (i.e. computations) of various kinds. They are not homogeneous in contents nor in length. At the end of the chapter we mention the relevant references.

VI.1 Some Simple Semigroups

Consider the following three Hamiltonian functions

$$H_1(p) = a \cdot p$$

$$H_2(p) = \frac{p^2}{2}$$

$$H_3(p) = \lambda \int (e^{\xi \cdot p} - 1)\, \mu\,(d\xi)$$

where $\mu(d\xi)$ is a measure on $\mathbb{R}^n$ such that the integrals above converge. The Hamilton–Jacobi functions corresponding to the $H(p)$ above are

$$S_i(q,P,t) = q \cdot P - t\, H_i(p) \qquad i = 1,2,3$$

and according to the results of chapter III we know that for a test function $\Phi(q)$, the function

$$u_i(q,t) = (T_{S(t)}\, \Phi)(q) \qquad i = 1,2,3$$

satisfies

$$\frac{\partial}{\partial t}\, u_i(q,t0) = H_i(-\nabla)\, u_i \qquad i = 1,2,3.$$

Let us compute the transition "densities" explicitly and then dualize. Begin with

$$u_1(q,t) = (T_{S(t)}\, \Phi)(q) = \int \exp - (iq \cdot k - it\, a \cdot k)\, \hat{\Phi}(k)\, \frac{dk}{(2\pi)^n}$$

$$= \Phi(q - at).$$

If μ denotes an element in the dual of the test functions then

$$\langle \mu, T_{S(t)} \Phi \rangle = \langle \mu, T_{-at} \Phi \rangle = \langle T_{at} \mu, \Phi \rangle$$

where T_{at} is the obvious translation operator. When $\mu = \varepsilon_{x_0}(dx)$ then $T_{at}\mu = \varepsilon_{x_0+at}(dx) = \delta[x - (x_0 + at)]\,dx$.

Consider now

$$u_2(a,t) = \left(T_{S_2(t)} \Phi\right) = \int \exp - i\left(q \cdot k + \frac{t\,k^2}{2}\right) \hat{\Phi(k)} \frac{dk}{(2\pi)^n}$$

$$= \int \frac{\exp - \frac{(q-\bar{q})^2}{2t}}{(2\pi t)^{n/2}} \hat{\Phi}(\bar{q})d\bar{q}$$

where we made use of the fact that the Fourier transform of $\dfrac{\left(\exp - \frac{q^2}{2t}\right)}{(2\pi t)^{n/2}}$ is $\exp - \dfrac{t\,k^2}{2}$.

The action of $T_{S_2(t)}$ on the dual space of the test functions is now easy to transport as well. For example, for any bounded function $f(q)$ we have that

$$U(q,t) \equiv P_t\, f(q) \equiv T_{S(t)}\, f(q) = \int f(\bar{q}) \exp - \frac{\left(q - \bar{q}\right)^2}{2t}.$$

This semigroup P_t corresponds to the standard Brownian motion on $\mathbb{R}^n$.

A simple extension of this example corresponds to a Hamiltonian $H(p) = \frac{1}{2}(p \,.\, A\, p)$ with A being a symmetric positive semidefiite matrix. In this way we obtain a process with generator

$$G = \frac{1}{2} \sum_{ij} a_{ij} \frac{\partial^2}{\partial q_i \partial q_j}.$$

We leave for the reader to obtain this case from the one above, when A is positive definite, by considering the canonical transformation generated by $P \,.\, A\, q = F(q,P)$.

The last case is the hardest ot compute explicitly in general. The best way to look at it is from the probabilistic corner. For that, let $X_1, X_2,\ldots,X_n,\ldots$ be a sequence of independent, identically distributed random variables with common distribution $\mu(d\xi) = P(X_i \in d\xi)$. Denote by $N(t)$ a Poisson process with parameter λ, independent of the $\{X_i\}$. We can now write for a test function Φ

$$u_3(q,t) = \int \exp \left\{ ik \,.\, q\, \lambda t \right\} [\exp (i\xi \,.\, k) - 1]\, \mu(d\xi) \Big\} \hat{\Phi}(k)\, dk$$

$$= \int \sum \frac{(\lambda t)^n}{n!} E\left[\exp (ik \,.\, X_1) \right[^n \exp (-\, ik \,.\, q)\, \hat{\Phi}(k) \frac{dk}{(2\pi)^n}$$

$$= \int E\left(E[\exp [ik \,.\, S(N(t))] \mid N(t)] \right) \exp (-\, ik \,.\, q)\, \hat{\Phi}(k) \frac{dk}{(2\pi)^n}$$

$$= E \int \exp [-\, ik \,.\, (q - Z(t))]\, \hat{\Phi}(k) \frac{dk}{(2\pi)^n}$$

$$= E\, \Phi(q) - Z(t)).$$

The missing steps are a standard computation for probabilists: expand $\exp t \int (\exp i\xi \,.\, k)\, \mu(d\xi)$ in powers of λt, identify $E[\exp i\xi \,.\, X_1]$ with $\int \exp i\xi \,.\, k\, \mu(d\xi)$.

For the rest, define $S(n) = X_1 + X_{2=} + \ldots + X_n$ and $S(N(t)) \equiv Z(t)$, and notice that $P(N(t)) = n = (\lambda t)^n \exp - \lambda t / n!$

Consider now any bounded f and notice that

$$<f, T_{S(t)}\, \Phi> = \int f(q)\, E\, \Phi(q - Z(t))\, dq = E \int f(q + Z(t))\, \Phi(q)\, dq$$

$$= \int E\, f(q + Z(t))\, \Phi(q)\, dq$$

and therefore

$$P_t\, f(q) = E\, [\, f(q + Z(t)\,]$$

$$= \sum \frac{(\lambda t)^n}{n!} \exp (-\, \lambda t)\, E\, [\, f(q + S(n)\,].$$

The distribution of the random variable $S(n)$ is the n-times convolution $\mu_n(dq) = \mu * \ldots * \mu\, (dq)$ of μ. One case is simple to compute: at every step you move a distance h with probability $p = 1$. In this case

$$P_t\, f = \sum_{n=0}^{\infty} \exp (-\, \lambda t)\, \frac{(\lambda t)^n}{m!} f(q + mh)$$

$$= \int \left\{ \sum_{m=0}^{\infty} \exp(-\, \lambda t) \frac{(\lambda t)^m}{m!}\, \delta(q' - q - mh) \right\} f(q')\, dq'.$$

Another important class of processes with a classical analogue are the symmetric stable processes on $\mathbb{R}^n$ of index α, $1 \leq \alpha \leq 2$. The classical Hamiltonian would be

$$H(p) = |p|^{\alpha} = \left(\sum p_i^2\right)^{\alpha/2}.$$

Beware: This $H(p)$ is not necessarily analytic in p.

For the sake of it, let us work out a couple of cases using the techniques of chapter V. That is, in each of the three cases we should compute

$$U_j(q,t) = f(C_j^+(t))\,1$$

where $C_j^+(t) = q + t\,v_j(D)$, $j = 1,2,3$. Recall identity (IV - 3.4)

$$\exp ik \,.\, (t\,V_j(D) + q) \exp i\bar{k} \,.\, q$$

$$= \exp t\,(H(ik + i\bar{k}) - H(i\bar{k})) \exp iq \,.\, (k + \bar{k}).$$

Recall that Fourier transforms in the dual of the test functions have the opposite sign conventions and assume f can be retrieved from its transform to obtain

$$f(C^+(t))\,1(q) = f(q + ta)$$

in the first case. For the second we have

$$f(C^+(t))\,1(q) = \int \hat{f}(k) \exp(ik \,.\, C^+(t)) \frac{dk}{(2\pi)^n}$$

$$= \int \exp(ik \,.\, q) - \exp(-k^2\,t/2)\,\hat{f}(k) \frac{dk}{(2\pi)^n}$$

$$= \int f(q') \exp - (q - q')^2 / 2t \;\frac{dk}{(2\pi)^{n/2}}$$

as above.

VI.2 A Particle in a Constant Field

We shall treat here the one dimensional case which is the general case anyway. Consider the Hamiltonian $H(q,p) = p^2/2 - Eq$ defined on $\mathbb{R}^2$.

Here we shall verify that the approaches introduced in chapters III and V coincide.

To begin note that

$$S(q,P,t) = q(P + Et) - \frac{Et^2}{2}P - \frac{t}{2}P^2 - \frac{E^2t^3}{6}$$

generates the canonical transformation mapping the coordinates (q,p) at time t onto the initial data at $t = 0$.

The transformations equations are

$$q = Q + Pt + \frac{Et^2}{2}$$

$$p = P + Et$$

and it is easy to verify that $\frac{\partial S}{\partial t} + \frac{1}{2}\left(\frac{\partial S}{\partial q}\right) - Eq = 0$. This transformation may be obtained by composing $q\,.\,P - \frac{P^2t}{2}$ with the transformation generated by $F(q,P) = qP + Et^2\frac{P}{2} + Eqt - \frac{Et}{6}$ which transforms (q,p) into $(Q,P) = (q + \frac{Et}{2}, p + Et)$, i.e., the accelerating particle is transformed into a free particle by jumping onto an accelerating frame. In other words: compensate the field away.

Anyway, let Φ be a test function and consider

$$u(q,t) = (T_{S(t)}\,\Phi)(q) = \int \exp - S(qw,ik,t)\,\hat{\Phi}(k)\,\frac{dk}{(2\pi)^n}$$

$$= \int \hat{\Phi}(k)\exp\left\{-ik\left(q - \frac{Et^2}{2}\right) - \frac{tk^2}{2} + \frac{E^2t^3}{6} - qEt\right\}\frac{dk}{(2\pi)^n}$$

$$= \int \Phi(\bar{q})\exp\left\{-\frac{(q-\bar{q})^2}{2t} - \frac{(q+\bar{q})}{2}Et + \frac{E^2t^3}{24}\right\}\frac{d\bar{q}}{\sqrt{2\pi t}}.$$

It is easy to verify that by dualization we obtain

$$U(q,t) = (P_t f) = \int f(\bar{q}) \exp\left\{ -\frac{(q-\bar{q})}{2t} - \frac{(q+\bar{q})}{2} Et + \frac{E^2 t^3}{24} \right\} \frac{dk}{(2\pi)^{n/2}}$$

which satisfies

$$\frac{\partial u}{\partial t} = \left(\frac{1}{2}\Delta - Eq\right) u.$$

To obtain $U(q,t) = f(C^+(t))\, \Omega_0(q)$ we note that $C^+(t) = q + \frac{Et^2}{2} + tD$, $C(t) = \nabla + Et$. To find $\Omega_0(q)$ take Fourier transforms of $\left(\frac{D^2}{2} - Eq\right)\Omega_0(q) = 0$ to obtain $\left(-\frac{k^2}{2} - iE\frac{d}{dk}\right)\hat{\Omega}_0(k) = 0$ which can be integrated to yield $\hat{\Omega}_0(k) = \exp -\frac{ik^3}{6E}$, giving

$$\Omega_0(q) = \frac{1}{2\pi}\int \exp(ikq)\, \hat{\Omega}_0(k)\, dk = \frac{1}{\pi}\int_0^\infty \cos\left(kq + \frac{k^3}{6E}\right) dk.$$

The function $\Omega_0(q)$ is well known; it is an Airy function. We leave for the reader to verify that

$$P_t\, \Omega_0(q) = \Omega_0(q)$$

for P_t given by (2.1). Let us compute $\Omega_0(q)^{-1} f(C^+(H))\ \Omega_0(q)$, for $f(q) = \int \exp ik \,.\, q \frac{\hat{f}(k)}{(2\pi)^n} dk$.

Note to begin with that

$$\exp ik \,.\, C^+(t) \exp i\bar{k}q = \exp\left\{ik\frac{Et^2}{2} - \frac{t(k+\bar{k})}{2} + \frac{k^2}{2}t + i(k+\bar{k})q\right\}$$

$$= \exp\left\{\frac{i(k+\bar{k})Et^2}{2} - \frac{t(k+\bar{k})^2}{2} + i(k+\bar{k})q + \frac{\bar{k}}{2}t - i\bar{k}\frac{Et^2}{2}\right\}$$

and note also that

$$\frac{i(k-iEt)^3}{6E} - \frac{E^2t^3}{6} = \frac{ik^3}{6E} - \frac{\bar{k}t}{2} - i\bar{k}\frac{Et^2}{2}$$

With this we obtain

$$f(C^+(t))\,\Omega_0(q)$$

$$\int\int \hat{f}(k) \exp\left\{ i(k+\bar{k})\left(q+\frac{Et^2}{2}\right) - \frac{(k+\bar{k})^2 t}{2} \exp\left\{ i\frac{(\bar{k}-iEt)}{6E} + \frac{Et}{6}\right\} d\bar{k}\, \frac{dk}{(2\pi)^2}$$

$$= \int d\bar{k} \int \hat{f}\,(k-\bar{k}) \exp\left\{ ik\left(q+\frac{Et^2}{2}\right) - \frac{k^2 t}{2}\right\} \exp\left\{ i\,\frac{(\bar{k}-iEt)^3}{6E} + E^2 t^3\right\} \frac{d\bar{k}}{(2\pi)^2}$$

$$= \int d\bar{k} \int \rho_t\left(q+\frac{Et^2}{2} - q'\right) \exp i\,\bar{k}q'\, f(q')\, dq' \exp\left\{ i(\bar{k} - tEt)^3 + \frac{E^2 t^3}{6}\right\} \frac{d\bar{k}}{(2\pi)}$$

where we put $\rho_t(q) = \dfrac{\left(\exp - \dfrac{q^2}{2t}\right)}{\sqrt{2\pi t}}$. Now

$$\int \exp i\,\bar{k}\,q' \exp \frac{i\left(\bar{k} - iEt\right)^3}{6E} \frac{d\bar{k}}{(2\pi)} = \exp - q'Et\,\Omega_0(q')$$

with which

$$\Omega_0(q)^{-1} f(C^+(t))\,\Omega_0(q) = \Omega_0^{-1}(q)$$

$$= \Omega_0^{-1}(q) \int \exp\left\{ - q'\,Et + \frac{E^2 t^3}{6}\right\} \rho_t\left(q + \frac{Et^2}{2} - q'\right) \Omega_0(q')\, f(q')\, dq'$$

$$= \Omega_0^{-1}(q)\, P_t\, f\, \Omega_0(q)$$

from which the kernel for P_t can be obtained and correlated with that of (2.1).

Observe that the method worked even though $\Omega_0(q)$ is not positive.

VI.3 The Oscillator and the Ornstein-Uhlenbeck Processes

In this and the following two sections we shall play with some quadratic but important Hamiltonians.

We shall begin by separating variables in the Hamiltonian

$$H = \frac{1}{2}\sum \Gamma_{ij}\, p_i\, p_j - \frac{1}{2}\sum V_{ij}\, q_i\, q_j\,. \tag{3.1}$$

by means of canonical transformations. We do this because the one dimensional case contains all that is needed and because we shall need the computation when we do linear filtering. In (3.1) V, Γ are symmetric, positive definite matrices.

Let D be an orthogonal matrix such that $D\,\Gamma\,D^+ = \bar{\Gamma}$ is diagonal and consider the canonical transformation generated by $F(q,\bar{p}) = q\,.\,D^+\,\bar{p} = Dq\,.\,\bar{p}$. This maps (q,p) onto $(\bar{q},\bar{p}) = (Dq,Dp)$ and H onto

$$\bar{H} = \frac{1}{2}\left(\sum \Gamma_i\, \bar{p}_i^2 - \sum_{ij} \bar{V}_{ij}\, \bar{q}_i\, q_j\right)$$

where of course Γ_i are the elements along the diagonal of Γ and $\bar{V} = D\,V\,D^+$. Let now $F_2(\bar{q}, \hat{p}) = \bar{q}\,.\,R\hat{p}$, with $R = \sqrt{\bar{\Gamma}}$, be the generating transformation mapping $(\bar{q},\bar{p})$ onto $(\hat{q},\hat{p}) = (Rq,Rp)$ and $\bar{V}$ into $\hat{V} = R^{-1}\,\bar{V}\,R^{-1}$. Now $\bar{H}$ becomes

$$\hat{H} = \frac{1}{2}\left(\sum_i \hat{p}_i^2 - \sum_{ij} \hat{V}_{ij}\, \hat{q}_i\, \hat{q}_j\right)$$

To finish consider $\hat{D}$ be the orthogonal matrix that diagonalizes $\hat{V}$. As in the first step by $F_3(\hat{q},P) = \hat{q}\,.\,\hat{D}^+ P$ we obtain coordinates $(Q,P) = (D\hat{q}, D\hat{p})$ and $\hat{H}$ becomes

$$\tilde{H} = \frac{1}{2}\sum_i \left(P_i^2 - w_i^2\, Q_i^2\right).$$

We shall denote our variables by (q,p) again and note to begin with that

$$\frac{1}{2}\left(\frac{d^2}{dq^2} - w^2\, q^2\right)\Omega_0(q) = -\frac{w}{2}\Omega_0(q)$$

with $\Omega_0(q) = \exp -\frac{q^2 w}{2}$. Thus we should consider the one dimensional Hamiltonian

$$H(q) = \frac{1}{2}(p^2 - w^2 q^2) + \frac{w}{2}$$

and its associated generator

$$G(q) = \frac{1}{2}\left(\frac{d^2}{dq^2} - w^2q^2\right) + \frac{w}{2}$$

satisfying $G\,\Omega_0(q) = 0$. We are after the density of the semigorup $P_t f(q)$ that satisfies $\frac{\partial P_t f}{\partial t} = G\,P_t\,f$. In chapter V we saw that

$$\Omega_0(q)^{-1}\,(P_t\,f\,\Omega_0)(q) = \Omega_0(q)^{-1}\,f(C^+(t))\,\Omega(q) = \tilde{P}_t\,f(q)$$

from which ρ_t can be obtained from $\tilde{\rho}_t$ according to comments after (V-1.5).

The classical equations of motion for $H(q,p)$ are $\dot{q} = p$, $\dot{p} = W^2\,q$ the solutions to which are

$$q(t) = q\cosh tw + \frac{p}{w}\sinh tw, \quad p(t) = wq\sinh tw + p\cosh tw$$

from which

$$C^+(t) = q\cosh tw + \frac{\sinh tw}{w}D, \quad C(t) = wq\sinh tw + \cosh tw\,D$$

for in this case theorem (2.11) applies neatly. If we compute, and this one is just as above, $f(C^+(t))\,\Omega_0(q)$, we obtain

$$f(C^+(t))\,\Omega_0(q)$$

$$= \int \hat{f}\,\exp - \left(\frac{k^2 \sinh tw\cosh tw}{2w}\right)\exp\,(ikq\cosh tw)$$

$$\Omega_0\left(q + k\,ik\,\frac{\sinh tw}{w}\right)\frac{dk}{(2\pi)}$$

Expanding the exponential in $\Omega_0(q)$, regrouping terms and putting $\sigma(t) = \frac{(1 - e^{-2wt})}{2w}$ we obtain

$$\Omega_0(q)^{-1}\,f(C^+(t))\,\Omega_0(q) = \int f(q')\exp - \frac{(q' - q\,e^{-wt})^2}{2\sigma(t)}\,\frac{dq'}{\sqrt{2\pi\sigma(t)}}$$

$$= \int f(q')\,\tilde{\rho}_t(q,q')\,dq'\,.$$

And therefore

$$\rho_t(q,q') = \Omega_0(q)\,\tilde{\rho}_t(q,q')\,\Omega_0(q)^{-1}$$

$$= \frac{1}{\sqrt{2\pi\sigma(t)}} \exp - \left[\frac{(q-q')^2 (1+e^{-2wt}) + 2qq'(1-e^{-wt})^2}{4\sigma(t)} \right].$$

It is easy to verify that both $\tilde{\rho}_t$ and ρ_t tend to the transition kernel of Brownian motion as $w \to 0$.

Let us examine the effect of the vacuum for state on $G = \frac{1}{2}\left(\frac{d^2}{dq^2} - w^2 q^2 \right) + \frac{w}{2}$. We saw that $\tilde{G} = \Omega_0(q)^{-1}\, G\, \Omega_0$, or

$$\tilde{G}\, f(q) = \frac{1}{2} \frac{d^2}{dq^2} f - w\, q \frac{d}{dq} f$$

which is the infinitesimal generator of the Ornstein-Uhlenbeck process. Notice that $\tilde{\Omega}(q) = 1$ is a vacuum for $\tilde{G}$.

Consider the Hamiltonian $\bar{H} = p^2/2 - wqp$. The corresponding Hamiltonian equations have solutions $q(t) = qe^{-wt} + \sinh \frac{tw}{w}$, which determine the operators

$$\bar{C}^{+}(t) = q\, e^{-tw} + \frac{\sinh tw}{w} D, \quad C(t) = e^{wt} D$$

and we leave for the reader to check that

$$f(\bar{C}(t))\, 1(q) = \tilde{P}_t\, f(q) = \Omega_0(q)^{-1}\, (P_t\, \Omega_0\, f)(q).$$

We should also remark that although $T_{S(t)}\ \Phi(q)$ is not a semigroup with $S(q,P,t)$ being given by

$$S(q,P,t) = \frac{qP}{\cosh wt} - \frac{\operatorname{tgh} tw}{w} \left(\frac{P^2}{2} - \frac{wq^2}{2} \right)$$

we can still use it to get G acting on test functions according to definition (IV-1.5). The results is (surprise!)

$$G\Phi = \frac{\partial}{\partial t}(T_{S(t)}\,\Phi)(q)\,|_{\,t=0} = \frac{1}{2}\Big(\frac{d^2}{dq^2} - \frac{q^2w^2}{2}\Big)\,\Phi(w).$$

We leave for the reader to work out the whole thing for the semigroup having as classical analogue of a particle subject to Newton's equation $\ddot{q} + \dot{q} + w^2 q = f(t)$, that is for the forced, damped, harmonic oscillator. This will be done using transforms while we do the linear filtering problem.

VI.4 A Particle in a Constant Electromagnetic Field

This problem has received a lot of attention in the physical literature. Consider a system moving in a 3-dimensional configuration space with Hamiltonian

$$H = \frac{1}{2}(p - A(q))^2 + E\,.\,q - \frac{\gamma^2}{2}q^2 \tag{4.1}$$

where $A(q) = Mq$, M being the matrix

$$M = w\begin{pmatrix} 0 & -1 & 0 \\ 1 & 0 & 0 \\ 0 & 0 & 0 \end{pmatrix}$$

E being a constant vector and, w, γ real constants. The matrix M has been chosen so that for any vector V, $MV = \frac{1}{2}\,B \wedge V$, with $B = \text{curl } A(q)$ and $\wedge$ denoting the standard vector product in $\mathbb{R}^3$.

Here we proceed to reduce the problem to that treated in the prveious section. Let $R(t) = \exp tM$. By means of

$$F_1(q,\bar{p},t) = R(t)\,q\,.\,\bar{p}$$

transform (4.1) into

$$\bar{H} = \frac{1}{2}\Big(\,\bar{p}^2 - q\,.\,\sigma^2\,q\,\Big) + E(t)\,.\,q \tag{4.2}$$

with $E(t) = R(t)E$ and σ^2 being the diagonal matrix with entries $\sigma_1^2 = \sigma_2^2 = w^2 + \gamma^2$, $\sigma_3^2 = \gamma^2$.

Let now $\xi(t)$ and $\eta(t)$ denote two time dependent vectors, such that

$$F_2(\bar{q},\hat{p},t) = \bar{q}\,.\,\hat{p} + \hat{p}\,.\,\xi(t) - \bar{q}\,.\,\eta(t) + \varphi(t)$$

ransforms $\bar{H}$ given by (4.2) into

$$\hat{H} = \bar{H} + \frac{\partial F_2}{\partial t} = \frac{1}{2}\left(\hat{p}^2 - (\hat{q} \,.\, \sigma^2 \hat{q})\right). \tag{4.3}$$

It is easy to verify that

$$\dot{\xi} = \eta, \quad \dot{\eta} = \sigma^2\xi + E(t), \qquad \dot{\varphi} + \frac{1}{2}\left(\dot{\xi} + \xi \,.\, \sigma^2\xi\right) = 0.$$

These equations can be integrated with zero initial conditions. We already now how to find the infinitesimal generator associated to (4.3). It is

$$G \equiv \sum_{i=1}^{3} G_i \equiv \sum_{i=1}^{3} \frac{1}{2}\left[\frac{d^2}{dq_i^2} - \sigma_i^2 q_i^2\right]$$

and now the whole analysis of section 3 can be applied.

VI.5 Other Quadratic and Integrable Hamiltonians

We present some oddities about processes associated to quadratic Hamiltonians which can be understood when canonical transformations play a role.

Notice to begin with that the Hamiltonians $H = -\frac{qp}{2}$ and $\tilde{H} = \frac{1}{2}(P^2 - QP)$ are related to each other by the canonical transformations generated by $F(a,P) = q \,.\, P + \frac{P^2}{2}$ and its inverse $\bar{F}(Q,p) = Qp - \frac{p^2}{2}$.

The equation

$$\frac{\partial u}{\partial t}(q,t) = -q\,\frac{\partial u}{\partial q} \tag{5.1}$$

generates the semigroup $P_t\, f(q) = f\,(q\, e^{-t/2})$ as is easy to see integrating (5.1) using characteristics, or as we did in chapter V: define $C^+(t) = q \exp -\frac{t}{2}\nabla_q$, then

$$P_t\, f(q) = f(C^+(t))\, 1(q) = f\left(q\, e^{-t/2}\right).$$

When we dualize, P_t acts on test functions $\Phi(q)$ as

$$P_t\,\Phi(q) = e^{t/2}\,\Phi(q\,e^{t/2}) \tag{5.2}$$

which is the semigroup generated by $G\,\Phi(q) = \dfrac{1}{2}\dfrac{\partial}{\partial q}(q\,\Phi)(q)$. Let us now use canonical transforms to find the semigroup $\tilde{P}$ on $\Phi(Q)$ given by $\tilde{P}_t = T_{\bar{F}}\,P_t\,T_F$.

Let $\Phi(Q)$ be a test function so that $\hat{\Phi}(k)$ vanishes outside a compact set and verify that

$$\left(T_F\,\Phi\right)^{\wedge}(k) = \exp\frac{k^2}{2}\hat{\Phi}(k)$$

and note that according to (5.2)

$$P_t\left(T_F\,\Phi\right)(q) = \exp\frac{t}{2}\left(T_F\,\Phi\right)(q\,e^{t/2})$$

whcih has Fourier transform

$$\left[P_t\,T_F\,\Phi\right]^{\wedge}(k) = \exp\left[\frac{k^2 e^{-t}}{2}\right]\hat{\Phi}\left(k\,e^{-t/2}\right)$$

and therefore

$$\tilde{P}_t\,\Phi(Q) = \left(T_{\bar{F}}\,P_t\,T_F\,\Phi\right)(Q)$$

$$= \int \exp(-ik\,.\,Q)\exp\left(-\frac{k^2}{2}\right)(P_t\,T_F\,\Phi)^{\wedge}(k)\,\frac{dk}{(2\pi)}$$

$$= \int \exp -ik\,.\,Q\,\exp -\frac{k^2}{2}(1 - e^{-t})\,\hat{\Phi}\,(k\,e^{-t/2})\,\frac{dk}{(2\pi)}$$

$$= \int \exp\left[\frac{-(Q - Q')^2}{2\sigma(t)}\right]\Phi(e^{t/2}\,Q')\,\frac{dQ'}{\sqrt{2\pi\sigma(t)}}$$

where $\sigma(t) = (1 - e^{-t})$. This is our old $\sigma(t)$ with $w = \dfrac{1}{2}$. One more change of variables gives

$$\tilde{P}\,\Phi(Q) = \int \exp\left[-\frac{(Q - Q'e^{-t/2})^2}{2\sigma(t)}\right]\Phi(Q')\,\frac{dQ'}{\sqrt{2\pi\sigma(t)}}$$

and when we dualize to obtain the action of $\tilde{P}_t$ on functions $f(Q)$ we get

$$\tilde{P}_t f(Q) = \int \frac{\exp\left[\dfrac{-(Qe^{-t/2} - Q')^2}{2\sigma(t)}\right](Q')dQ'}{\sqrt{2\pi\sigma(t)}} \tag{5.3}$$

which is the semigroup of the Ornstein-Uhlenbeck process with generator

$$\tilde{G} = \frac{1}{2}\left(\frac{\partial^2}{\partial Q^2} - Q\frac{\partial}{\partial Q}\right).$$

Actually, the representation given by theorem (V-2.11) is valid for any Hamiltonian which is a quadratic function of (Q,p). Combining this fact with Williamsons classification of quadratic Hamiltonians (see appendix 7 of [V-5]) we can readily obtain all semigroups (and processes) associated to quadratic Hamiltonians.

We shall consider two simple cases obtainable from quadratic systems by means of canonical transformations. When analyzing the classical Heisenberg-Weyl algebras we shall consider another case.

Consider the canonical transformation generated by $F(q,P) = \varphi(q) \cdot P$ with $\varphi(q)$ a diffeomorphism of $\mathbb{R}^n$. Let us begin with a Hamiltonian $H(p) = p^2/2$ which after transformation becomes

$$\tilde{H}(Q,P) = \sum h_{ij}(Q) P_i P_j \quad \text{with} \quad h_{ij} = \sum \frac{\partial \varphi_i}{\partial q_k}(\psi(Q)) \frac{\partial \varphi_j}{\partial q_k}(\psi(Q))$$

where $\psi(Q)$ is the inverse to $\varphi(q)$.

If we wanted to solve $\partial\tilde{u}/\partial t = \tilde{G}\,\hat{u}$ with $\hat{u}(0,Q) = \tilde{f}(Q)$ and

$$\tilde{G}\,\tilde{u} = \sum \frac{\partial \varphi_k}{\partial q_i} \frac{\partial}{\partial Q_k}\left(\frac{\partial \varphi_j}{\partial q_i} \frac{\partial \tilde{u}}{\partial Q_j}\right)$$

the obvious changes of variables can be expressed as $\tilde{u}(t) = \tilde{P}_t \tilde{f} = T_F P_t T_F^{-1} f, f = T_F \tilde{f}$, where P_t is the semigroup with generator $G = \sum \dfrac{\partial^2}{\partial q_i^2}$.

Had we started with the Hamiltonian $H = \sum q_i p_i$ we would have obtained upon transformation the Hamiltonian

$$\tilde{H}(Q,P) = \sum \psi_i(Q) \frac{\partial \varphi_j}{\partial q_i}\big(\psi(Q)\big) P_j$$

Therefore the problems

$$\frac{\partial u}{\partial t} = \sum q_i \frac{\partial u}{\partial q_i}, \quad u(0,q) = f(q)$$

and

$$\frac{\partial \tilde{u}}{\partial t} = \sum \psi_i(Q) \frac{\partial \varphi_j}{\partial q_i}(\psi(Q)) \frac{\partial \tilde{u}}{\partial Q_j}, \quad \tilde{u}(0,Q) = (T_F^{-1} f)(Q)$$

are canonically conjugate.

A variation on this theme appears when $H = \sum_{i=1}^{n} X_i(q)\, p_i$ and $G = \sum X_i(q) \frac{\partial}{\partial q_i}$ with $X(q) = (X_1(q),\ldots,X_n(q))$ being a globally integrable field on $\mathbb{R}^n$.

If we write $q_i(t) = \bar{q}_i(t,q)$ for the solution to $\dot{q}_i = X_i(q)$ passing through q at $t = 0$ then $C_1^+(t) = \bar{q}_i(t,C^+(0))$ and since the vacuum for G is $\Omega_0 \equiv 1$ we can write the solution t

$$\frac{\partial u}{\partial t} = \sum X_i(q) \frac{\partial u}{\partial q_i}, \quad u(0,q) = f(q)$$

as

$$u(t,q) = f(C^+(t))\, \Omega_0(q) = f(\bar{q}(t,q))$$

which is obvious from the method of characteristics.

Consider now linear perturbations of integrable systems, i.e., Hamiltonians of the form $H(q,p) = K(p) + a \,.\, q$. The solution to $\dot{q}_i = v_i(p) \equiv \frac{\partial K}{\partial p_i}$ and $\dot{p}_i = -a_i$ now being $p(t) = p - a$ and therefore $\dot{q}_i = v_i(p - at)$ has $q_i(t) = q + c_i(p,t)$ as solution where

$$c_i(p,t) = \int_0^t v_i(p - as)\, ds = \frac{\partial}{\partial p_i} \int_0^t K(p - as)\, ds.$$

The vacuum state for $G = K(\nabla) + a \,.\, q$ is to be found solving $(K(\nabla) + a \,.\, q)\, \Omega_0(q) =$ or taking Fourier transforms

$$(K(ik) - ia \,.\, \nabla_k)\, \hat{\Omega}_0(k) = 0 \quad \text{with} \quad \hat{\Omega}_0(k) = \int \exp(-ik \,.\, q)\, \Omega_0(q) dq$$

hich can be solved by rotating the coordinate system to place the n-axis (in k-space) ong q, solving a 1-dimensional equation and undoing the rotation. Assuming that done, e recover $\Omega_0(q)$ by

$$\Omega_0(q) = \int \hat{\Omega}_0(k) \exp ik \cdot q \frac{dk}{(2\pi)^n}.$$

With this the solution to

$$\frac{\partial u}{\partial t} = (K(\nabla) + a \cdot q) u \qquad u(0,q) = f(q)$$

an be found by first solving

$$\frac{\partial \tilde{u}}{\partial t} = \tilde{G} \tilde{u}, \quad \tilde{G} = \Omega_0^{-1} G \Omega_0 \quad \tilde{u}(0,q) = f(q)$$

e solution to which is given by

$$\tilde{u}(t,q) = \Omega_0(q)^{-1} f(C^+(t)) \Omega_0(q)$$

hich can be computed writing $\hat{f}$ and $\Omega_0(q)$ in terms of their Fourier transforms and using roposition (IV.33).

A variant of this example consists in perturbing the particle moving with constant speed y an arbitrary potential, i.e., by considering $H(q,p) = v \cdot p + V(q)$ with v being a constant ector. This Hamiltonian is obtainable from the former by means of the canonical ansformation $(Q,P) = (-p,q)$ which is not obtainable from a generating function of the pe $F(q,P)$.

In this case, once we have $\Omega_0(q)$ satisfying $(v \cdot \nabla + V(q)) \Omega_0(q) = 0$, the gauge ansform of $G = v \cdot \nabla + V(q)$ is $\tilde{G} = v \cdot \nabla$ and $\partial\tilde{u}/\partial t = v \cdot \nabla\tilde{u}$ solved by

$$\tilde{u}(t,q) = \tilde{f}(q + vt) \equiv \tilde{P}_t f(q)$$

nd solution to

$$\frac{\partial u}{\partial t} = (v \cdot \nabla + V(q)) u \qquad u(0,q) = f(q)$$

given by

$$u(t,q) = P_t f(q) = \Omega_0(q) \frac{f(q+vt)}{\Omega_0(q+vt)}$$

as can be verified by direct computation.

VI.6 The Linear and Non Linear Filtering Problems

Filtering theory is concerned with obtaining the best prediction for (a function of) the state of the system given a signal or observation process. The problem being that both the system and the observations may be under the action of random influences.

The simplest model, known as the linear filtering problem is described by

$$dx_i = \sum a_{ij}\, dw_j$$

(6.

$$dy_i = \sum c_{ij}\, x_j\, dt + dv_i$$

where $w_j(t)$, $v_i(t)$ are independent Brownian motions, each of zero mean and variance t. To describe the problem in intuitive terms: x_i represents the velocity of a particle, determined by random forces. We want to know $x(t)$ but we can only observe some linear function $y(t)$ of the position. And the measurement of $y(t)$ is also contaminated by noise.

It can be shown that the unnormalized conditional density of $x(t)$ given $\{y(s), s \leq$ satisfies the Duncan-Mortensen-Zakai equation

$$\frac{\partial \rho}{\partial t} = \left\{ \frac{1}{2} \sum_{ijk} a_{ik}\, a_{jk} \frac{\partial^2}{\partial q_i \partial q_j} - \frac{1}{2} \sum_{ijk} c_{ki}\, c_{kj}\, q_i\, q_j \right\} \rho + \sum \xi_i\, x_i\, \rho \equiv G.$$

Here $\xi_i = \sum c_{ik} \frac{dv_k}{dt}$ (which and $dy_k\,(dt)$ should be formally understood as Stratonovitch derivative. (This refers to the integration rules one has to apply to $\xi(t)$).

Consider the associated classical random Hamiltonian

$$H(q,p) = \frac{1}{2}\{p\,.Ap - q\,.\,Vq\} + \xi\,.\,q \qquad (6.$$

where $A = a\,a^+$ and $V = C^+ C$ are two $n \times n$-matrices and we shall assume a to be non-singular.

Let us begin by standardizing the Hamiltonian (6.3) by means of the canonical transformation generated by

$$F_1(q,\bar{p}) = (\bar{p}, a^{-1} q)$$

which transforms (6.3) into

$$\bar{H}(\bar{q},\bar{p}) = \frac{1}{2}\{\bar{p}^2 - (\bar{q} \cdot \bar{V}\bar{q})\} + \bar{\xi} \cdot \bar{q} \tag{6.4}$$

with $\bar{V} = a V a^+$ and $\bar{\xi} = a^+ \xi$. Let now D be an orthogonal matrix transforming $\bar{V}$ into diagonal form W with entries $w_1^2 \delta_{ij}$ and consider the transformation generated by $F_2(\bar{q},\hat{p}) =$ $\hat{p} \cdot D\bar{q}$. Now $\bar{H}(\bar{q},\bar{p})$ becomes

$$\hat{H}_2(\bar{q},\hat{p}) = \frac{1}{2} \sum \hat{p}_i^2 - w_i^2 q_i^2 + \hat{\xi} \cdot \hat{q} \tag{6.5}$$

with $\hat{\xi} = D\bar{\xi}$.

We can now eliminate the linear term in (6.5) by means of the transformation generated by

$$F_3(\hat{q},P) = \hat{q} \cdot P + f \cdot P - \dot{f} \cdot \hat{q} + \varphi(t)$$

which is specified by the change of variables

$$\hat{Q} = \hat{q} + f, \qquad P = \hat{p} + \dot{f}$$

and the new Hamiltonian being

$$\tilde{H}(Q,P) = \hat{H} + \frac{\partial F}{\partial t} = \frac{1}{2}(P^2 - Q \cdot WQ) = \frac{1}{2} \sum (P_i^2 - w_i^2 Q_i^2) \tag{6.6}$$

if f and φ are chosen so that

$$\ddot{f} - \Omega f = \hat{\xi}, \quad \dot{\varphi} + \frac{1}{2}(\dot{f}, \dot{f}) + \frac{1}{2}(f, \Omega f) = 0$$

with zero initial conditions.

The infinitesimal generator associated to $\tilde{H}$ is

$$\tilde{G} = \frac{1}{2} \sum_{i=1}^{n} \left(\frac{\partial^2}{\partial Q_i^2} - W_i^2 Q_i^2 \right) \equiv \tilde{G}_i . \tag{6.7}$$

Observe now that all the randomness present in (6.2) and (6.3) has disappeared for both $\tilde{H}$ and $\tilde{G}$ are perfectly deterministic objects. The randomness has been shifted to the generating function $F_3(\hat{q},P)$ and to its associated operator T_{F_3}.

We leave for the reader to write the solution to (6.2) for any initial data ρ by applyin all the transforms involved. The same problem was done in [VI-3] by percolating it throug the quantum mechanical formulation.

Consider a variation on the same theme, but this time for a one-dimensional filterin problem which is Itô form looks like

$$dx_t = f(x_t)\,dt + g(x_t)\,dw$$

$$dy_t = h(x_t) + dv_t\,.$$

Here f, g, h are such that all computations below make sense. In this case th unnormalized density $\rho(q,t)$ of $x(t)$ given $\{y_s : s \leq t\}$ satisfies the DMZ-equation

$$\frac{\partial\rho}{\partial t} = \frac{1}{2}\frac{\partial^2}{\partial q^2}(g^2\rho) - \frac{\partial}{\partial q}f\rho - \frac{1}{2}h^2(q)\rho + h(x)\xi\rho \qquad (6.8$$

where again, ξ is to be treated formally as the Stratonovitch derivative dv/dt.

Rewrite (5.8) as

$$\frac{\partial\rho}{\partial t} = \frac{1}{2}g\frac{\partial}{\partial q}\left(g\frac{\partial}{\partial q}\right)\rho - M(q)\frac{\partial\rho}{\partial q} - \frac{1}{2}V(q)\rho + h(q)\xi\rho = G\rho \qquad (6.8$$

where $M(q) = f - \frac{1}{2}g,\, g'$, $V = 2f' - (g^2)'' + h^2$ and as usual $\frac{d}{dq}$ is denoted by a prime symbo

Put

$$\varphi(q) = Q = \int_0^q \frac{dq'}{g(q')}$$

and consider to begin with the transformation generated by

$$F_1(q,P) = P\,\varphi(q)$$

which changes (6.8) or (6.8)′ into

$$\frac{\partial\rho}{\partial t} = \frac{1}{2}\frac{\partial^2}{\partial q^2}\tilde{\rho} - \tilde{M}(Q)\frac{\partial\tilde{\rho}}{\partial q} - \frac{1}{2}U(Q) + \tilde{h}(Q)\,\xi\tilde{\rho} \qquad (6.9$$

here $\tilde{\rho}(Q,t) = (T_F\, \rho)(Q) = \rho(\varphi^{-1}(Q),t)$ and $\tilde{M}(Q) = M(\varphi^{-1}(Q))$, $V(Q) = V(\varphi^{-1}(Q))$, $(Q) = h(\varphi^{-1}(Q))$. Let us now apply the gauge transformation generated by

$$F(Q,\hat{p}) = Q\,\hat{p} + A(Q)$$

ith $A'(Q) = \tilde{M}(Q)$. Denote the new canonical coordinate also by Q. Then (6.9) becomes

$$\frac{\partial \hat{\rho}}{\partial t} = \frac{1}{2}\,\frac{\partial^2}{\partial Q^2}\hat{\rho} - W(Q)\,\hat{\rho} + \hat{h}(Q)\,\xi(t)\,\hat{\rho} \tag{6.10}$$

where $\hat{\rho}(Q,t) = \exp A(Q)\,\tilde{\rho}(Q,t)$ and $W(Q) = \frac{1}{2}\left(M'(Q) - M^2(Q) + U(Q)\right)$.

It is clear at this stage that this problem reduces to the one treated above whenever $V(Q) = \frac{1}{2}(aQ^2 + bQ + c)$, and we can apply the methods we developed for the linear iltering problem.

Certainly this sequence of transformations consists of simple changes of variables, but onsistency demands that they fit our general scheme.

VI.7 Canonical Transforms and Umbral Calculus

Here we redo and motivate better the results in [VI.5]. The main goal of this section is to provide a scheme within which to relate different polynomial sequences among themselves.

We are going to be considering polynomials of binomial type, Scheffer sequences and cross sequences for the time being. Later on we shall relate them to moment systems. The former three are usually expressed in terms of their generating functions by

$$g(q,\xi) = \exp q\,.\,U(\xi) = \sum_m \xi^m\, P_m(q)\,/\,m! \tag{7.1}$$

What do we get by shifting our point of view thus? Well, note that one gets all possible $\exp q\,.\,U(ik)$ by varying U (or V) in $T^*_V\, e_k(q)$, and using the fact that $U \to T_U$ is a representation of the group structure we obtain on $\{U : \mathbb{R}^n \to \mathbb{R}^n$, invertible, analytic$\}$ by taking composition as the group law. Also in the set of trivially integrable Hamiltonians we have the orbit structure induced by the diffeomorphisms U. The orbit of H is given by $O(H) = \{H(U) : U : \mathbb{R}^n \to \mathbb{R}^n$, analytic and integrable$\}$ and we go from orbit to orbit as in proposition (III–3.6).

In other words we have only one basic polynomial sequence of binomial type and one basic cross-sequence. The rest are obtainable from these by canonical transformations.

Before we simplify notations, let us draw a few diagrams that should be in the back of everybody's mind. If we denote by F_Q, F_q, F'_Q, F'_q our spaces of test functions in the Q-(or q) variables and their respective duals we have the diagrams

Here we denote by Σ (or $\tilde{\Sigma}$) any of $q \, . \, P \pm H(P)t$ $\left(q \, . \, P \pm t\tilde{H}(P)\right)$.

From now on we drop the $*$ in $T_U^* \equiv T_F^*$, let $F(q,P) = q \, . \, U(P)$ and note that in the distribution sense

$$T_V f(q) = \int \hat{f}(k) \exp(q \, . \, U(ik)) \frac{dk}{(2\pi)^n} \tag{7.4}$$

Let us begin our development of umbral calculus with two lemmas that are a particular case of what we did in chapters III and IV.

Assume throughout that $U(0) = V(0) = 0$.

Lemma 7.5 *Let* $X : \mathbb{R} \to \mathbb{R}$ *be such that* $X(D)$ *is defined then*

$$X(V(D))T_V = T_V X(D).$$

Lemma 7.6 $T_V f(0) = f(0)$ *for every* $f(Q) \in F'_Q$.

Proofs (7.6) is for the reader. To get (7.5) notice that

$$T_V X(D) e_k = X(ik) T_V e_k = X(ik) \exp(q \, . \, U(ik))$$

$$= X(V(D)) \exp(q \, . \, U(ik)) = X(D) T_V e_k.$$

Let us now verify that the T_V given in (7.4) actually correspond to the series representation of *umbral operators*.

For that expand the exponential in (7.4) to obtain

$$T_V f(q) = \int \exp(q \,.\, U(ik))\, \hat{f}(k) \frac{dk}{(2\pi)^n}$$

$$= \sum_m \frac{q^m}{m!} \int [\, U(D)^m f\,]^\wedge (k) \frac{dk}{(2\pi)^n}$$

$$= \sum_m \frac{q^m}{m!} [\, U(D) f\,](0)$$

where here $D = \nabla_Q$, but there is no need to distinguish variables.

The expansion (7.7) is usually termed the *umbral expansion* of f.

Definition 7.8 The polynomials $b_m(Q) = Q^m$ will be called the *basic polynomials.*

Lemma 7.9 $T_V\, b_m(q) = P_m(q)$

Proof

First proof: expand $\exp ik \,.\, Q = \sum (ik)^m \frac{b_m(Q)}{m}$ and apply T_V to obtain

$$T_V\, e_k = \sum \frac{(ik)^m}{m!} T_V\, b_m(q) = \exp q \,.\, U(ik) = \sum \frac{(ik)^m}{m!} P_m(q).$$

Second proof:

$$(T_V\, b_m)(q) = \int \exp(q \,.\, U(ik))\, \hat{b}_m(k) \frac{dk}{(2\pi)^n}$$

$$= \int \exp(q \,.\, U(ik))\, (-\, i\, \nabla_k)^m\, \delta(k) \frac{dk}{(2\pi)^n}$$

$$= \int [(\nabla_{ik})^m \exp(q \,.\, U(ik)\,]\, \delta(k) \frac{dk}{(2\pi)^n}$$

$$= \sum_{l} \left(\frac{P_l(q)}{l} \right) \int [(\nabla_{ik})^{\mathbf{m}} (ik)^{\mathbf{l}}] \, \delta(k) \frac{dk}{(2\pi)^n}$$

$$= P_{\mathbf{m}}(q).$$

Let U_1, U_2 be as above and V_1, V_2 be their inverses and consider $U_2(U_1)$ and $V_1(V_2)$. Then $T_{V_1 \circ V_2} = T_{V_2} T_{V_1}$ and denoting by $P^{(i)}_{\mathbf{m}}$ be the polynomial sequences

$$P^{(i)}_{\mathbf{m}} = T_{V_i} b_{\mathbf{m}}$$

then

Lemma 7.10 $P^{(2)}_{\mathbf{m}}(q) = T_{U_1 \circ V_2} P^{(1)}_{\mathbf{m}}$.

Proof From (7.9) $b_{\mathbf{m}} = T^{-1}_{V_1} P^{(1)}_{\mathbf{m}} = T_{U_1} P^{(1)}_{\mathbf{m}}$, therefore

$$P^2_{\mathbf{m}}(q) = T_{V_2} b_{\mathbf{m}}(q)$$

$$= T_{V_2} T_{U_1} P^{(1)}_{\mathbf{m}} = T_{U_1 \circ V_2} P^{(1)}_{\mathbf{m}}.$$

Comment If we denote by $C(m,k)$ the coefficients of $T_{U_1 \circ V_2} b_m(q)$, and by $a^{(i)}(\mathbf{m},\mathbf{k})$ those $P^{(i)}_{\mathbf{m}}(q)$ we get

$$a^{(2)}(\mathbf{m},\mathbf{k}) = \sum_{\mathbf{l} \le \mathbf{k} \le \mathbf{m}} a^{(1)}(\mathbf{m},\mathbf{k}) \, C(\mathbf{k},\mathbf{l}) \tag{7.11}$$

and the ordering $\mathbf{k} \ge \mathbf{l}$ means the obvious thing: coefficient-wise ordering. Implicit in (7.11) is the fact that $C(\mathbf{k},\mathbf{l})$ vanishes when $\mathbf{l} \not\le \mathbf{k}$, that is the binomial polynomials $P_{\mathbf{m}}(q)$ are of degree $|\mathbf{m}| = \sum m_i$.

Since both $b_{\mathbf{m}}(q)$ and $P_{\mathbf{m}}(q)$ can be used as a basis for polynomial expansions, before generalizing Taylor's expansion formula we need

Lemma 7.12 *For any multi-index* $\mathbf{m}$

$$\left[D^{\mathbf{m}} T_U f \right](0) = \left[V(D)^{\mathbf{m}} f \right](0).$$

Proof From (7.5) $X(V(D))T_V = T_V X(D)$ we obtain if we pre- and postmultiply by T_U

$$T_U X(V(D)) = X(D)T_U$$

Now we apply to f in F_q' and use lemma (2.6) to obtain

$$\left[T_U X(V(D))f\right](0) = \left[X(D)\, T_U f\right](0) = [\, X(V(D))\, f\,](0)$$

which is what we want.

Lemma 7.13 $T_V\, f(q) = \sum_m \frac{1}{m!} P_m(q) \left(D^m f\right)(0).$

Proof It is a variation on the theme of the second proof of (7.9). Just expand the exponential in (7.4) as given by (7.1) and use some Fourier calculus.

Proposition 7.14 (The Generalized Taylor Expansion). *Let f have a power series expansion. Then*

$$f(q) = \sum \frac{1}{m!}\; P_m(q) \left[V(D)^m f\right](0).$$

Proof

$$f(q) = T_V\, T_U\, f(q) = \sum \frac{1}{m!}\; P_m(q) \left[D^m\, T_U f\right](0)$$

$$= \sum \frac{1}{m!}\; P_m(q) \left[V(D)^m f\right](0)$$

where the second identity is lemma 7.13 with f replaced by $T_U f$. The last follows from (7.13) with $X(D) = D^m$.

Comment If we denote by $\bar{P}_m$ the polynomial sequence $T_U\, b_m$, then

$$b_m(q) = \sum_{k \le m} P_k(q) \left(D^k \bar{P}_m\right)(0).$$

The range of the sum is restricted to $\mathbf{k \le m}$ since $\left(D^k \bar{P}_m\right)(0)$ vanishes for $\mathbf{k \le m}$.

To better understand our version of the Rodrigues, Steffensen and Gzul formulae recall the effect of $F(q,P) = q \cdot U(P)$. Under this transformation $P_i = V_i(q)$ and

$$Q_i = \sum q_j \frac{\partial U_j}{\partial P_i}(V(q)),$$

the identities

$$b_{\mathbf{m}+\mathbf{e}_i}(Q) = Q_1\, b_{\mathbf{m}}(Q), \qquad \frac{\partial}{\partial Q_i} b_{\mathbf{m}}(Q) = m_i\, b_{\mathbf{m}-\mathbf{e}_i}(Q)$$

are the result of applying the operators $\hat{Q}_i$ and $\hat{P}_i$ to $b_{\mathbf{m}}$. Here $\mathbf{e}_i$ is the multi-index with all entries equal to zero except the i-th which equals 1.

Proposition 7.15 *With the notations introduced above*

$$P_{\mathbf{m}+\mathbf{e}_i}(q) = \sum q_j \frac{\partial U_j}{\partial P_i}(V(D))\, P_{\mathbf{m}}(q).$$

Proof We challenge the reader to do it using the results of chapter IV. Let's do transforms

$$P_{\mathbf{m}+\mathbf{e}_j}(q) = \int \exp\,(q \cdot U(ik)) \left[q_j \cdot b_{\mathbf{m}} \right]^{\wedge}(k) \frac{dk}{(2\pi)^n}$$

$$= \int \exp\,(q \cdot U(ik))\, i \frac{\partial}{\partial k_j} \left[b_m \right]^{\wedge}(k) \frac{dk}{(2\pi)^n}$$

$$= \int \left(\frac{\partial}{\partial i k_j} \exp q \cdot U(ik) \right) \left[b_m \right]^{\wedge}(k) \frac{dk}{(2\pi)^n}$$

$$= \sum q_\ell \frac{\partial U_\ell}{\partial P_j}(V(D)) \int \exp\,(q \cdot U(ik)) \left[b_m \right]^{\wedge}(k) \frac{dk}{(2\pi)^n}$$

$$= \sum q_\ell \frac{\partial U_\ell}{\partial P_j}(V(D))\, P_{\mathbf{m}}(q).$$

Comment This is a Rodrigues type identity.

Proposition 7.16 (The G-Formula). *If* $U = V^{-1}$, $V(\xi) = \xi - G(\xi)$ *and* G *and its final derivatives vanish at* $\xi = 0$, *then*

$$T_V = \sum J_V(D) \frac{G^m(D)}{m!} \hat{b}_m(\cdot)$$

where $J_V(\xi)$ *denotes the Jacobian determinant of* V *and* $\hat{b}(\cdot)$ *is the obvious multiplication operator.*

Proof Make the change of variables ik into V(ik) in $T_U f(q)$ to obtain

$$(T_U f)(q) = \int \exp(q \, . \, U(ik)) \, \hat{\Phi}(k) \frac{dk}{(2\pi)^n}$$

$$= \int J_V(ik) \exp(iq \, . \, k) \, \hat{f}\left(\frac{V(ik)}{i}\right) \frac{dk}{(2\pi)^n}$$

$$= J(D) \int \exp(ik \, . \, q) \, \hat{\Phi}\left(\frac{V(ik)}{i}\right) \frac{dk}{(2\pi)^n}.$$

Consider now $\hat{f}\left(\frac{V(ik)}{i}\right)$.

$$\hat{f}\left(\frac{V(ik)}{i}\right) = \int \exp(- Q \, . \, V(ik)) \, \tilde{f}(Q) \, dQ$$

$$= \int \exp(- iQ \, . \, k + Q \, . \, G(ik)) \, f(Q) \, dQ$$

$$= \sum_m \frac{G(ik)^m}{m!} \int \exp(- iQ \, . \, k) \, Q^m f(Q) \, dQ$$

$$\sum_m \frac{G(ik)^m}{m!} \left[b_m f\right]^\wedge (k)$$

which can then be inserted in the expression we got above for T_U to obtain the desired result.

The following two results are called the Steffensen identities of the first and second kind, respectively.

Proposition 7.17 *Put* $U = V^{-1}$ *and* $V = \xi - \xi H(\xi)$, H *analytic and* H(0) = 0. *If* $R(\xi) = (1 - H(\xi))^{-1}$, *then for any multi-index* **m**

$$P_{\mathbf{m}}(q) = J_V(D)\, R^{|\mathbf{m}|+n}(D)\, b_{\mathbf{m}}(q).$$

Proposition 7.18 *With the same assumptions as in* (7.17)

$$P_{\mathbf{m}+\mathbf{e}_j} = \sum x_\ell \left(\frac{\partial U_i}{\partial P_j}\right)(V(D))\, J_V(D)\, R^{|\mathbf{m}|+n}(D)\, b_{\mathbf{m}}(q)$$

Proofs A glance at (7.17) and (7.15) suffices to obtain (7.18). A proof of (7.17) using transforms goes as follows

$$P_{\mathbf{m}}(q) = \int \exp(q \,.\, U(ik)) \left[b_m\right]^{\wedge}(k) \frac{dk}{(2\pi)^n}$$

$$= J_V(D) \int \exp(iq \,.\, k) \left[b_m\right]^{\wedge}\left(\frac{V(ik)}{i}\right) \frac{dk}{(2\pi)^n}$$

and

$$\left[b_m\right]^{\wedge}\left(\frac{F(ik)}{i}\right) = \int \exp(-\,Q \,.\, V(ik))\, b_{\mathbf{m}}(Q)\, dQ$$

$$= \int \exp\left(\frac{-\,ik \,.\, Q}{R \,.(ik)}\right) b_{\mathbf{m}}(Q)\, dQ$$

$$= R(ik)^{|\mathbf{m}|+n} \int \exp(-\,ik \,.\, Q)\, b_m(Q)\, dQ = R(ik)^{|\mathbf{m}|+n}\left[b_{\mathbf{m}}\right]^{\wedge}(k)$$

and therefore

$$P_{\mathbf{m}}(q) = J_V(D) \int R(ik)^{|\mathbf{m}|+n} \exp(ik \,.\, q)\left[b_{\mathbf{m}}\right]^{\wedge} \frac{dk}{(2\pi)^n}$$

$$= J_V(D)\, R(D)^{|\mathbf{m}|+n}\, P_{\mathbf{m}}(q).$$

To present an application of these identities we present Garsia and Joni's derivation of the multidimensional Lagrange inversion formula.

To begin with, note that if $F(\xi) = \sum F_m \frac{\xi^{\mathbf{m}}}{\mathbf{m}!}$ then

$$F_{\mathbf{m}} = \left[\,[\,F(D)\,\xi^{\mathbf{m}}\,\right](0).$$

Let now $\Phi(\xi)$ be an analytic function of ξ and we want to obtain $\Phi(U(\xi))\,|_m$ with $U = V^{-1}$ as above. Then

$$\Phi(U(D))|_m = [\Phi(U(D))\, b_m](0) = [\Phi(U(D))T_U\, T_V\, b_m](0)$$

$$= [T_U\, \Phi(U(V(D)))\, T_V\, b_m](0) = [T_U\, \Phi(D)\, P_m](0) = [\Phi(D)\, P_m](0)$$

where we used the computation in lemma (7.12) for the third step, $T_V\, b_m = P_m$ in the fourth step and lemma (7.6) in the fifth step.

We leave for the reader to verify that

$$\Phi(U)|_m = \left(\Phi\, J_V\, T^{|m|+n}\right)_m .$$

VI.8 Moment Systems and Cross Sequences

Now, let us get moving. That is, let us combine the results in the previous section with time evolution. To recall some notation introduced in chapter V: X(t) and $\tilde{X}(t)$ will denote generalized processes with semigroups P_t and $\tilde{P}_t$ respectively, which are related by the gauge transformation $\tilde{P}_t\, \Omega_0^{-1}\, P\, \Omega$. We denote by Ω_0 the vacuum for P_t which satisfies $G\, \Omega_0(q) = 0$.

Let $C_i^+(t), C_j(t),\ 1 \le i,\ j \le n$ be the operators defined by

$$C_i^+(t) = \exp t[G, \cdot\,]\, C_i^+(0)$$

$$C_i(t) = \exp t[G, \cdot\,]\, C_i(0)$$

and introduce the operators

$$A_i^+(t) = \hat{\Omega}_0(q)^{-1}\, C^+(t)\, \hat{\Omega}_0(q)$$
$$A_j(t) = \hat{\Omega}_0(q)^{-1}\, C(t)\, \hat{\Omega}_0(q) \tag{8.1}$$

where again the caret indicates that the funciton is to be considered a multiplicative operator.

Definition 8.2 The moment system associated to (G,Ω_0) is defined by

$$h_0(q,t) \equiv 1$$
$$h_m(q,t) = (A^+(t))^m\, 1(q) = \Omega_0(q)^{-1}\, ((C^+(t))^m\, \Omega_0)(q). \tag{8.3}$$

We saw in chapter V, that the right hand side of (8.3) can be written as

$$\tilde{E}^q(\tilde{X}(t))^{\mathbf{m}} = \tilde{P}(\tilde{X}(t))^{\mathbf{m}} = \Omega_0(t)^{-1}\, E^q\,[\, X^{\mathbf{m}}(t)\, \Omega_0(X(t))\,]$$

which explains the reason for the name *moment system* assigned to $h_m(q,t)$.

Lemma 8.4 *The operators* $A_i^+(t), A_j(t),\ 1 \le i,\ j \le n,$ *satisfy*

$$[\, A_i(t), A_j^+(t)\,] = \delta_{ij}\,.$$

Lemma 8.5 *The following identities hold*

(a) $A_i^+ h_{\mathbf{m}} = h_{\mathbf{m}+\mathbf{e}_i}.$

(b) $A_i\, h_{\mathbf{m}} = m_i\, h_{\mathbf{m}-\mathbf{e}_i}.$

(c) $A_i^+ A_i\, h_{\mathbf{m}} = m_i\, h_{\mathbf{m}}.$

(d) $\dfrac{\partial}{\partial t} h_m = \tilde{G}\, h_{\mathbf{m}}.$

Proof Property (a) follows from the definition, and (b) follows from the definition and the fact that $\left[A_i, A_j^{m_j} \right] = m_j\, A^{m_j - 1}\, \delta_{ij}$ which follows trivially from lemma (8.4) or the time evolution operator $\exp t[G, \cdot\,]$ applied to $\left[\dfrac{\partial}{\partial q_i}, q_j^{m_j} \right] = m_j\, q^{m_j - 1}\, \delta_{ij}.$

Property (c) is a combination of (a) and (b), and (d) follows from the definition of $\tilde{A}^+(t)$ or from the fact that $h_{\mathbf{m}}(q,t) = P_t(\tilde{X}(t))^{\mathbf{m}}$.

Let us now compute the generating function of the sequence $h_{\mathbf{m}}(q,t)$

$$g_t(q,ik) \equiv \sum \frac{(ik)^{\mathbf{m}}}{\mathbf{m}!}\, h_m(q,t) = \Omega_0(q)^{-1}\, [\exp\,(-ik\,.\, C^+(t)\,\Omega)\,](q) \tag{8.6}$$

and if $\hat{f}(k)$ is the Fourier transform of f, we have

$$\int \hat{f}(k)\, g_t(q,ik) \frac{dk}{(2\pi)^n} = \Omega_0(q)^{-1} \left(\int \hat{f}(k) \exp\,[ik\,.\,C^+(t)\,] \frac{dk}{(2\pi)^n} \Omega_0 \right)(q)$$

$$= \tilde{P}_t\, f(q) = \int f(q)\, \rho_t(q, q')\, dq.$$

In other words we have

$$g_t(q,ik) = \int \tilde{\rho}_t(q,q') \exp(ik \,.\, q')\, dq'$$

$$= \tilde{E}^q [\exp(ik \,.\, \tilde{X}(t))].$$

Comments When $G = H(D)$, then $C^+(t) = q + tv(D)$, $C(t) = D$. In this case $\Omega_0(q) = 1$ and $P_t = \tilde{P}_t$, and we saw in the first section of this, chapter that

$$g_t(q,ik) = \exp ik \,.\, C^+(t)\, 1(q)$$
$$= E^q [\exp ik \,.\, X(t)] = \exp ik \,.\, q + t\, H(ik) \tag{8.8}$$

in other words, *for trivially integrable systems cross-sequences and moment systems are the same thing: compare* (8.8) *with* (7.3).

We are going to examine now how the umbral operators T_V introduced above relate moment sequences to one another.

Again it proves useful to distinguish functions of the Q–variables from functions of the q–variables.

Let us consider first the case in which $G1 = 0$.

Consider the Q_i as multiplication operators. Then after transforming by means of the T_V we obtain

$$\tilde{C}_i^+(0) = \sum C_j^+(0)\, J_{ji}(1(0)), \quad \tilde{C}_i(0) = V_i(C(0))$$

with $\tilde{C}_i^+(0)$ and $C_j(0)$ as above. We leave to the reader to verify that

$$\left[\tilde{C}_i(0), \tilde{C}_j^+(0)\right] = \delta_{ij} \tag{8.9}$$

applied to $T_V f$ with f in F'_Q. Consider now

$$\tilde{C}_i^+ = \exp t\,[G, \cdot\,]\, \tilde{C}_i^+(0), \quad \tilde{C}_i(0) = \exp t[G, \cdot\,]\, \tilde{C}_i(0)$$

which combined with (8.9) yields

$$\left[\tilde{C}_i(t), \tilde{C}_j^+(t)\right] = \delta_{ij}\,.$$

Define

$$h_m(q,t) = \left(\tilde{C}^+(t)\right)^m 1(q) \tag{8.10}$$

to which proposition (8.5) applies verbatim. We can state now

Proposition 8.11 *Let* $P_{\mathbf{m}}(q) = T_V\, b_{\mathbf{m}}(q)$. *Then the* $\tilde{h}_{\mathbf{m}}(q,t)$ *defined in* (8.10) *satisfy*

$$\tilde{h}_{\mathbf{m}}(q,t) = P_{\mathbf{m}}(C^{+}(t))\, 1(q)$$

for every multi-index $\mathbf{m}$.

Proof From the analogue of (8.5) we know that

$$\frac{\partial \tilde{h}_{\mathbf{m}}}{\partial t} = G\, \tilde{h}_{\mathbf{m}}(q,t)$$

and that at $t = 0$

$$\tilde{h}_{\mathbf{m}}(q,0) = (\tilde{C}^{+}(0))^{\mathbf{m}}\, 1(q) = P_{\mathbf{m}}(q) = P_{\mathbf{m}}\,(C^{+}(0))\, 1(q)$$

where the second identity is the Rogrigues-like formula (7.15). Consider now the sequence

$$h'_{\mathbf{m}}(q,t) = P_{\mathbf{m}}(C^{+}(t))\, 1(q).$$

Since $\exp t\, G1(q) = 1$ we have $P(C^{+}(t))1(q) = \exp tGP_{\mathbf{m}}(C^{+}(0))1(q) = \exp t\, G\, P_{\mathbf{m}}(q)$. Therefore $h'_{\mathbf{m}}(q,t)$ satisfies the same equation and the same initial condition as $\tilde{h}_{\mathbf{m}}$ Ergo, $h_{\mathbf{m}} = h'_{\mathbf{m}}$.

Comment What (8.110) means is the following. When the $h_{\mathbf{m}}(q,t)$ and the $P_{\mathbf{m}}(q)$ are known, the $\tilde{h}_{\mathbf{m}}(q,t)$ can be obtained as

$$\tilde{h}_{\mathbf{m}}(q,t) = \sum_{\mathbf{k}\leq\mathbf{m}} a(\mathbf{m},\mathbf{k})\, (C^{+}(t))^{\mathbf{k}}\, 1(q) = \sum_{\mathbf{k}\leq\mathbf{m}} a(\mathbf{m},\mathbf{k})\, h_{\mathbf{k}}(q,t).$$

We also have that $V_i(C(0))\, P_{\mathbf{m}}(q) = m_i\, P_{\mathbf{m}-e_i}\, 1(q)$ (which is obtained applying T_V to $\frac{\partial b_{\mathbf{m}}}{\partial} = m_i\, b_{\mathbf{m}-e_i}(q)$. If we let everything evolve in time, the analogue of (8.5-b) yields

$$V_i(C(t))\, P_{\mathbf{m}}\,(C^{+}(t))\, 1(q) = m_i\, P_{\mathbf{m}-e_i}\,(C^{+}(t))\, 1(q).$$

We consider now the case in which <u>$G\,1 \neq 0$</u> but we have our positive vacuum $\Omega_0(q)$.

The key ingredient above was the writing Q_i and P_i in terms of (q,p) to which we associated multiplicative and derivative operators respectively, and such that $P_i\, 1(q) = \frac{\partial 1}{\partial q^i} = 0.$

The corresponding objects in this case are

$$\hat{q}_i\, \Omega_0(q) = q_i\, \Omega_0(q), \qquad \hat{p}_i\, \Omega_0(q) = 0$$

which are to satisfy

$$(\hat{q})^m\, \Omega_0(q_i) = \Omega_0\, b_m(q)\, 1, \qquad (\hat{p})^m\, \Omega_0 = \Omega_0 (D)^m\, 1 = 0$$

and therefore we can choose

$$\hat{q}_i : \text{multiplication by } q_i$$

$$\hat{p}_i = \frac{\partial}{\partial q_i} - \frac{\partial}{\partial q_i} \ln \Omega_0 = \frac{\partial}{\partial q_i} - \frac{1}{\Omega_0} \frac{\partial \Omega_0}{\partial q_i}.$$

We challenge the reader to do this by means of the gauge transformation we introduced above.

We introduce now

$$R_i(0) = \sum_{l=1}^{n} \hat{q}_l\, J_{li}\,(\hat{p})\,, \qquad L_i(0) = V_i(\hat{p}) \tag{8.12}$$

and notice also that $L_i(0)\Omega_0 = 0$ since we assume analyticity of V.

Then

$$[\,L_i, R_j\,]\,\Omega_0 = \delta_{ij}\, \Omega_0 \tag{8.13}$$

and we define

$$\begin{aligned} R_i(t) &= \exp t[\,G, \cdot\,]\, R_i(0) \\ L_i(t) &= \exp t[G, \cdot\,]\, L_i(0) \end{aligned} \tag{8.14}$$

and the polynomial sequences

$$\bar{h}_m(q,t) = (R(t))^m\, \Omega_0(q) \tag{8.15}$$

and we have

Proposition 8.16 *With the notations introduced up to* (8.15) *we have*

(a) $(\tilde{P}_t R_m)(q) = \Omega_0(q)^{-1} (R(t))^m \Omega_0(q)$.

(b) $R_i(t) \tilde{h}_m(q) = \tilde{f}_{m+e_i}(q)$.

(c) $L_i(t) h_m(q) = m_i \tilde{f}_{m+e_i}(q)$.

(d) $\dfrac{\partial h_m}{\partial t} = G h_m$.

Proof (b), (c), (d) follow from the definitions. To obtain (a), do as follows

$$(R_t(t))^m \Omega_0(q) = P_t (R(0))^m \Omega_0(q)$$

$$= P_t \Omega_0 P_m(\hat{q}) 1(q) = P_t (P_m \Omega_0)(q).$$

Comment It takes only a glance to notice that (8.16) coincides with (8.11) when $\Omega_0(q) = 1$ Can you guess why we use the notation R and L from looking at (b) – (c) above?

Let us now consider trivially integrable systems. Let H(p) be an analytic function o P; and let us consider the monomials b_m to begin with.

We know that now $P_t = T_{S(t)}$ and we have already noticed that

$$b_m(q,t) \equiv T_{S(t)} (C^+(0))^m 1(q) = (q + t\, v(D))^m 1(q), \quad v_i(D) = \frac{\partial H}{\partial p_i}(D)$$

is such that

$$\sum \frac{(ik)^m}{m} b_m(q,t) = \exp ik\, C^+(t)\, 1(q) = \exp ik \,.\, q + tH(ik) \qquad (8.17$$

which can be rewritten as

$$P_t e_k(q) = T_{S(t)} e_k = T_{S(t)} \exp (ik \,.\, C^+(0))\, 1(q)$$

with $e_k(q) = \exp ik \,.\, q$. Had we started with another Hamiltonain $\tilde{H}(p)$ we would have obtained

$$\tilde{b}_{\mathbf{m}}(q,t) \equiv T_{\tilde{S}(t)}(C^{+}(0))^{\mathbf{m}}\,1(q) = (q + t\,\tilde{v}(0))^{\mathbf{m}}\,1(q),\ \ \tilde{v}_i(D) = \frac{\partial \tilde{H}}{\partial q_i}(D)$$

nd also instead of (8.17),

$$\sum \frac{(ik)^{\mathbf{m}}}{\mathbf{m}} \tilde{b}_{\mathbf{m}}(q,t) = \exp ik\, C^{+}(t)\, 1(q) = \exp ik\,.\,q + t\,\tilde{H}(ik). \tag{8.18}$$

We have seen in a variety of places how to relate (8.17) to (8.18)

$$T_{S(t)}\, e_k = T_{S(t)}\, T_{\Sigma(t)}\, T_{S(t)}\, e_k = T_{\Sigma(t)\circ S(t)}\, T_{S(t)}\, T_{S(t)}\, e_k$$

vith $\Sigma(q,p) = q\,.\,P + tH(P)$ being the inverse to $S(q,P) = q\,.\,P - t\,H(p)$. What we are sserting in the last chain is nothing but

$$\exp ik\,.\,q + t\,\tilde{H}(ik) = \exp\left[\tilde{H}(ik) - H(ik)\right]\exp ik\, b\, q + t\, H(ik)$$

vhich can be used to relate the $\tilde{b}_{\mathbf{m}}(q,t)$ to the $b_{\mathbf{m}}(q,t)$ as follows

$$\tilde{b}_{\mathbf{m}}(q,t) = \sum_{\mathbf{k}\le \mathbf{m}} \binom{\mathbf{m}}{\mathbf{k}} B(\mathbf{m}-\mathbf{k})\, b_{\mathbf{k}}(q,t$$

$$B(\mathbf{m}) = \sum_{\mathbf{k}\le \mathbf{m}} \binom{\mathbf{m}}{\mathbf{k}} \tilde{\mu}_{\mathbf{m}-\mathbf{k}}(-t) \tag{8.19}$$

nd

$$\exp t\,\tilde{H}(ik) = \sum (ik)^{\mathbf{m}}\, \frac{\tilde{\mu}_{\mathbf{m}}(t)}{\mathbf{m}!},\quad \exp - t\, H(ik)) = \sum (ik)^{\mathbf{m}}\, \frac{\mu_{\mathbf{m}}(-t)}{\mathbf{m}!}.$$

The passage from the $b_{\mathbf{m}}(q,t)$ to the $\tilde{b}_{\mathbf{m}}(q,t)$ depended only on the fact that at $t = 0$; $b_{\mathbf{m}}(q,0) = \tilde{b}_{\mathbf{m}}(q,0) = b_{\mathbf{m}}(q)$. When $\tilde{H} = H(U)$ for some analytic diffeomorphism U, there is nother way of expressing the connection made above.

Note that $T_V\, e_k(q) = \exp U(ik)\,.\,q$ and $T_{S(t)}\, T_V\, e_k = \exp U(ik)\,.\,q + tH(U(ik))$ and if ve apply T_U to this we obtain

$$T_U\, T_{S(t)}\, T_V\, e_k(q) = \exp ik \,.\, q + t\, H(U(ik))$$

$$= \exp ik \,.\, q + t\, \tilde{H}(ik) = T_{\tilde{S}(t)}\, e_k(q).$$

We should have distinguished the Q and q variables. Do it!.
What we did is to play with the diagram (7.4) above.

The intermediate step in the former computation is of importance for it gives us the generating function of a family of polynomials obtained by propagating (operator-wise) the image of the $b_{\mathbf{m}}(Q)$ by means of T_V. It is a particular case of what we did to obtain (8.14 and (8.16). However, in this case the structure of the $C^+(t)$ operators can be written down more explicitly. Let

$$C_i^+(0) = \sum q_j\, J_{ji}(\nabla_q), \qquad C_i(0) = V_i(\nabla_q)$$

then

$$C_i^+(t) = \sum_1 (q_j - v_j(D))\, J_{ji}(D), \quad C_i(t) = v_i(D)$$

and we have

$$h_{\mathbf{m}}(q,t) = T_{S(t)}\, P_{\mathbf{m}}(C^+(0))\; 1 = P_t\, P_{\mathbf{m}}(q) = E[\, P_{\mathbf{m}}(q + X(t))\,]$$

which have the generating function

$$T_{S(t)}\, T_V\, e_k(q) = \sum \frac{(ik)^m}{\mathbf{m}!} h_m(q,t)\;\; pU(k) \,.\, q + t\, H(ik)).$$

Let us now work out a couple of simple cases. Let us begin with the Hamiltonian resulting in the oscillator process: $H = \dfrac{(p^2 - w^2 q^2)}{2}$. In this case the equations of motion have the solution

$$q(t) = q \cosh wt + p \sinh \frac{hwt}{w}.$$

$$p(t) = wq \sinh wt + p \cosh wt$$

which yields

$$C^+(t) = C^+(0)\cosh wt + \sinh wt \frac{C(0)}{w},$$

$$C(t) = w \sinh wt\, C^+(0) + \cosh wt\, C(0).$$

We know that the vacuum for $G = \frac{(D^2 - w^2 q^2)}{2}$ is $\Omega_0(q) = \exp - \frac{q^2 w}{2}$. Here we put $D \equiv \frac{\partial}{\partial q}$. Therefore the $A^+(t)$ and $A(t)$ operators are given by

$$A^+(t) = C^+(t) - \sinh wt\, C^+(0), \qquad A(t) = C(t) - w \cosh wt\, C^+(0).$$

The moment "polynomials" have generating function given by (8.6) or (8.7)

$$g(q,ik) = \int \tilde{\rho}_t(q,q') \exp ik\,.\,q'\, dq'$$

where $\tilde{\rho}_t(q,q')$ was obtained in section 3 and is given by

$$\tilde{\rho}_t(q,q') = \frac{1}{\sqrt{2\pi\sigma(t)}} \exp - \frac{(2' - qe^{-wt})}{2\sigma(t)}$$

with $\sigma(t) = \frac{(1 - e^{-2wt})}{2w}$. Computing the characteristic function we obtain

$$g_t(q,ik) = \exp\left(ikqe^{-wt} - \frac{k^2}{2}\sigma(t)\right)$$

and if we make use of the well known expansion

$$\exp\left(ik\xi - \frac{k^2}{2}\right) = \sum (ik)^n \frac{h_n(\xi)}{n!}$$

where the $h_n(x)$ are (up to a $\sqrt{2}$ scale factor) the Hermite polynomials, we then obtain

$$g_t(q,ik) = \sum (ik)^n \frac{h_n(q,t)}{n!}$$

with

$$h_n(q,t) = \sqrt{\sigma(t)^n}\, h_n\left(\frac{q\,e^{-wt}}{\sqrt{\sigma(t)}}\right).$$

We should mention in passing that the usual creation and annihilation operators are in this case

$$\alpha^+ = \frac{(wA^+(t) - A)}{\sqrt{2w}}, \quad \alpha = \frac{(wA^+(t) + A)}{\sqrt{2w}}$$

which again satisfy

$$[\alpha, \alpha^+] = [A, A^+] = 1$$

from which we readily obtain that the collection

$$\psi_n(q) = (\alpha^+)^n \, 1(q)$$

satisfies

$$\psi_{n+1} = \alpha^+ \psi_n, \quad \alpha\psi_n = n\psi_{n-1}$$

and a simple computation shows that

$$\alpha^+ \alpha = \frac{\left\{A^2 - w^2 A^{+2} - w[A, A^+]\right\}}{2w}$$

or

$$\tilde{G} = w\,\alpha^+ \alpha + \frac{w}{2}$$

and therefore

$$\tilde{G}\psi_n = w\left(n + \frac{1}{2}\right)\psi_n$$

which is the result in elementary quantum mechanics books.

Let us now consider the Hamiltonian $H(q,p) = aq + V(q)$ (with both q and p in $\mathbb{R}$) leading to $G = aD + V(q)$ having $\Omega_0(q) = \exp - \int V(\xi)\, d\xi$ as vacuum state.

Now $\dot{q} = 1, \dot{p} = - V'(q)$ and therefore

$$C^+(t) = C^+(0) + at, \quad C(t) = C(0) + \frac{(V(C^+(0)) - V(C^+(0) + at))}{a}.$$

this case the moment polynomials are

$$h_n(q,t) = \Omega_0(q)^{-1}\, C^+(t)^n(q) = \Omega_0(q+at)^n$$

hich clearly shows the effect of the perturbation $V(q)$.

Consider now the case of $H(q,p) = K(q) + aq$ (q,p in $\mathbb{R}$) which leads to $G =$ (D) + aq. Now the equations of motion are $\dot{q} = v(p) = K'(p)$, $\dot{p} = -q$. We now obtain

$$C^+(t) = C^+(0) + \frac{(K(C(0))-K(C(0)+aT))}{a}, \quad C(t) = C(0) - at.$$

As the vacuum for $G = K(D) + aq$ we take

$$\Omega_0(q) = \int \frac{\exp\left(i\xi q - T(i\xi)/a\right)d\xi}{2\pi N(a)} \tag{2.20}$$

here we assume that the normalization function $N(a)$ exists such that $\frac{\left(\exp - iT(\xi)/a\right)}{2\pi N(a)}$ onverges to $\delta(\xi)$ as a tends to zero. The generating function of the moment system

$$h_n(q,t) = \Omega_0(q)^{-1}\, C^+(t)^n\, \Omega_0(q)$$

$$g_t(q,ik) = \Omega_0(q)^{-1} \exp ik\, C^+(t)\, \Omega_0(q).$$

This can be computed using the exponential formula contained in (IV-3.3), but instead e shall solve

$$\frac{\partial g_t}{\partial t} = \tilde{G}\, g_t, \qquad g_0(q,ik) = \exp(ikq) \tag{8.21}$$

y first solving

$$\frac{\partial \bar{g}_t}{\partial t} = G\, \bar{g}_t, \qquad \tilde{g}_0(q,ik) = \exp(ikq) \tag{8.22}$$

nd then obtain $\bar{g}_t$ from g_t as indicated in the first section of chapter V. By Fourier trans- orming (8.22) we obtain

$$\frac{\partial \hat{\bar{g}}_t}{\partial t}(\xi,ik) = (K(i\xi) - iaD_\xi)\, \hat{\bar{g}}_t(\xi,ik), \quad \hat{\bar{g}}_0(\xi,ik) = 2\pi\delta(k-\xi)$$

or

$$\left(\frac{\partial}{\partial t} + iaD_\xi\right)\hat{\bar{g}}_t = K(i\xi)\,\hat{g}_t$$

which, looked along the flow $\Phi_\tau(t_0, \xi_0) = (t_0 + \tau, \xi_0 + ia\tau)$, can be rewritten as

$$\frac{d}{d\tau}\hat{\bar{g}}_t\,(\varphi_\tau(t_0, \xi_0)) = K(i\xi_0 + ia\tau))\,\hat{g}(\Phi_\tau(t_0, \xi_0))$$

which can be integrated to

$$\hat{\bar{g}}_t(\xi,ik) = 2\pi\delta(\xi - iat - k)\exp\int_0^t K(i\xi + as)\,ds$$

and undoing the Fourier transform we obtain

$$\bar{g}_t(q,ik) = \int \delta(\xi - iat - k)\exp\int_0^t K(i\xi + as)\,ds\,\exp(i\xi q)\,d\xi$$

$$= \exp ik(q + iat)\exp\int_0^t K(i(k + iat) + as)\,ds \tag{8.23}$$

$$= \exp ik(q + iat)\exp\int_0^t K(ik - as)\,ds$$

after the change of variable $t - s$ into s is performed once more. We also see that when $a = 0$ we re-obtain the solution to equation (8.22) with $G = K(\nabla)$.

Again denote $\exp ikx$ by $e_k(x)$ and observe that

$$g_t(q,ik) = \tilde{P}_t\,e_k(q) = \Omega_0(q)^{-1}\,P_t(e_k\,\Omega_0)(q)$$

$$= \int \hat{\Omega}_0(\xi)\,(P_t\,e_{k+\xi})\,(q)\,\frac{d\xi}{(2\pi)}.$$

But

$$P_t\,e_{k+\xi}\,(q) = \bar{g}_t(q,i(k + \xi))$$

and therefore

$$\bar{g}_t(q,ik) = \Omega_0(q)^{-1} \int \hat{\Omega}_0(\xi)\, \bar{g}_t(q,i(k+\xi))\, \frac{d\xi}{2\pi}$$

nd again if we expand $\bar{g}_t(q,i(k+\xi))$ given by (8.23) and substitute in the expression above, e obtain the $h_n(q,t)$. When a tends to 0, $\hat{\Omega}(\xi)$ tends to $2\pi\delta(\xi)$, $\Omega_0(q)$ to 1 and $\bar{g}_t(q,ik)$ $\exp ikq + tH(ik)$ as it should be

I.9 A Lie Algebra and its Representation

n this section we shall consider the algebra of inhomogeneous polynomials of order two, (q,p) where (q,p) are in $\mathbb{R}^2$, with the bracket multiplication

$$[X,Y] = \frac{\partial X}{\partial q}\frac{\partial Y}{\partial p} - \frac{\partial X}{\partial p}\frac{\partial Y}{\partial q}$$

As we pointed out at the end of chapter II, with $X(q,p)$ we can associate a family of anonical transformations with generating functions $W(q,P,s)$ satisfying

$$\frac{\partial W}{\partial s} + X\left(q, \frac{\partial W}{\partial q}\right) = 0 \quad W(q,P,0) = q \,.\, P \tag{9.1}$$

here $X(q,p) = b\,p^2 + a\,q\,p + c\,q^2 + d\,p + e\,q + m$. From $W(q,P,s)$ we shall find the perator X_{op} associated to X, and its vacuum state, after which we shall point out a onnection with some of the recent work by Feinsilver, [VI–8].

We shall leave for the reader to verify that by means of the canonical transformation enerated by $\alpha q P + (\gamma P^2 - \beta q^2)/2$ we can change $X(q,p) = b\,p^2 + a\,q\,p + c\,q^2 + d\,p +$ $q + m$ into

$$\bar{X}(Q,P) = k\,P\,Q + \bar{c}\,Q + \bar{d}\,P + m.$$

We should mention that we either preassign k and find the α, β, γ that do the job or et $\alpha = 1$ and find the k, β, γ that do the job. Once this is done, change the notations and eplace $\bar{X}(Q,P)$ by

$$X(q,p) = a\,q\,p + b\,p + c\,q + m. \tag{9.2}$$

In order to solve (9.1) for X given by (9.2) it is easier to perform one more canonical ransformation generated by $F(q,P) = q \,.\, P + \beta P - \gamma q$ and reduce $X(q,p)$ to $\bar{X}(Q,P) =$ $Q\,P - \bar{m}$ with $\bar{m} = m - b\,c\,/\,q$. It is now real easy to verify that $\gamma = c/a$, $\beta = b/a$, as vell as that the inverse to $F(q,P)$ is

$$\bar{F}(Q,p) = Q\,p - \beta\,p + \gamma\,q - \beta\,\gamma.$$

Now (9.1) becomes

$$\frac{\partial \overline{W}}{\partial s} + a\,Q\,\frac{\partial W}{\partial Q} + \bar{m} = 0$$

which can be easily solved by the method of characteristics yielding

$$\bar{W}(Q,p,s) = Q \exp(-as)\,p - m\,s.$$

If we did some diagram chasing, we would obtain

$$W(q,P,s) = F \circ \bar{W} \circ \bar{F}(q,P,s)$$

$$= q\,P \exp(-as) - \frac{b}{a}[\,1 - \exp(-as)\,]\,p - \frac{c}{a}[\,1 - \exp(-as)\,]\,q - mt \tag{9.3}$$

$$+ \frac{bc}{a}\int_0^s [\,1 - \exp(-at)\,]\,dt.$$

Another way to solve (9.1) would be to consider the flow of the Hamiltonian $X(q,p) = a\,q\,p + b\,p + c\,q + m$, i.e., to solve

$$\frac{dq(s)}{ds} = \frac{\partial X}{\partial p'}, \qquad \frac{dp(s)}{ds} = -\frac{\partial X}{\partial q'}, \quad (q(0), p(0)) = (Q,P)$$

and use the transformation equations

$$\frac{\partial W}{\partial q} = p, \qquad \frac{\partial W}{\partial P} = Q$$

to obtain $W(q,P,s)$.

Observe that when a tends to zero, $W(q,P,s)$ tends to

$$W_0(q,P,s) = q\,.\,P - bsp - csq - mt + \frac{b\,c}{2}t^2. \tag{9.4}$$

When we apply $T_{W(s)}$ to a function $f(q)$, according to (III-9.3) we should obtain

$$(T_{W(s)}\,f)(q) = \exp\{c\,\Delta(s)q + ms - cb\,\delta(s)\}\,f(q\exp(-as) + b\,\Delta(s)) \tag{9.5}$$

where $\Delta(s) = [1 - \exp(-as)]/a$ and $\delta(s) = \int_0^s \Delta(t)\,dt$. When a tends to zero, (9.6) tends to

$$(T_{W_0} f)(q) = \exp\{csq + ms - \frac{cb}{2}s^2\}\, f(q + bs) \qquad (9.6)$$

as it should.

According to definition (IV-1.5) we have

$$X_{op}(q,p)\, f(q) = \frac{\partial}{\partial s}(T_{W(s)} f)(q) = (aq\frac{\partial}{\partial q} + b\frac{d}{dq} + cq + m)\, f(q) \qquad (9.7)$$

which when α tends to zero becomes

$$X_{op}(q,p)(q) = \left(b\frac{d}{dq} + cq + m\right) f(q) \qquad (9.8)$$

as would follow from (9.6). For what follows, in (9.5) and (9.6) the value of s stays fixed, so we shall set $s \equiv 1$.

If we denote the $W_0(q,P)$ in (9.4) by $W_0(b,c,m)$, it follows (it is easier to use 9.6) that

$$W_0(b_1,c_1,m_1) \circ W_0(b_2,c_2,m_2) = W\left(b_1 + b_2, c_1 + c_2, m_1 + m_2 - \frac{(c_1 b_2 - b_2 c_1)}{2}\right) \qquad (9.6)$$

from which it follows that the inverse to $W_0(b,c,m)$ (for the composition law, of course) is

$$W_0(b,c,m)^{-1} = W_0(-b, -c, -m).$$

Since the inhomogeneous first degree polynomials in (p,q), i.e. the class $X(q,p) = bp + cq + m$ forms a Lie algebra under the Poisson bracket product (a realization of the Heisenberg-Weyl or quantum algebra) what we obtain by means of (9.6) or (9.9) is a group generated by such algebra.

In the first case, we obtain a group of integral operators and in the second case we obtain a group of canonical transformations.

If we keep a fixed and write $W(b,c,m)$ for the $W(q,P,s = 1)$ given by (9.3) we obtain still another representation of the Heisenberg-Weyl group, reducing to the ones described above when a tends to zero. The result bears a vague resemblance to (9.3) but is a bit more cumbersome to write down.

Recall that if we wanted to solve

$$\frac{\partial U}{\partial t} = X_{op}\, u \qquad u(q,0) = f(q)$$

by the techniques developed in chapter V we would have to start by finding a vacuum fc X_{op}. As above it is easier to solve

$$\tilde{X}_{op}\, \tilde{\Omega}_0(Q) = \alpha Q \frac{\partial}{\partial Q} \tilde{\Omega}_0 - \bar{m}\ \Omega_0 = 0$$

and then to obtain $\Omega_0(q)$ by

$$\Omega_0(q) = \left(T_F\, \tilde{\Omega}_0\right)(q).$$

Proceeding, we obtain $\tilde{\Omega}_0(Q) = Q^{\bar{m}/\alpha}$ from which we obtain $\Omega_0(q) = (\exp - cq/a$ $(q + d/m)^{\bar{m}/\alpha}$ which is readily verified to satisfy

$$\left[aq \frac{\partial}{\partial q} + cq + d \frac{\partial}{\partial q} + m \right] \Omega_0(q) = 0. \qquad (9.10$$

The semigroup P_t with generator X_{op} is given by

$$P_t\, f(q) = \exp\left\{ \frac{e}{a} [\, 1 - \exp(-\,at)\,] + mt + \frac{cd}{a^2} [\, 1 - at - \exp\,(-\,at)\,] \right\}$$

$$f\left(q \exp\,(at) + \frac{d}{a} [\exp\,(at) - 1]\right)$$

which is $T_{W(t)}\, f(q)$. See (9.5). Let us play with 9.10 for a while. Rename the coordinate as Q are rewrite (9.10) as

$$\left(\alpha Q \frac{\partial}{\partial Q} + cq + d \frac{\partial}{\partial Q} \right) \qquad (9.10'$$

where m was set equal to $-\lambda$. Apply to $\Omega_0(Q)$ sides the operator T_F with $F(q,P) =$ $q \,.\, P - TP^2/2$ mapping (Q,P) onto $(q,p) = (Q + P, P)$. Then (9.10)′ becomes

$$\left(aT \frac{\partial^2}{\partial q^2} + a\, q \frac{\partial}{\partial q} + T(c + d) \frac{\partial}{\partial q} + cq \right) \Omega_0(q,T) = \Lambda\, \Omega_0(q,T) \qquad (9.11$$

here

$$\Omega_0(q,T) = (T_F\,\Omega_0)(q) = \int \frac{\exp - (q-Q)^2}{\sqrt{2\pi T}}\,\Omega_0(Q)\,dQ$$

$$= \exp - \frac{1}{2T}\left[\frac{Tc}{a}q + 4\left(\frac{Tc}{a}\right)\right]$$

$$= \int \frac{\exp - Q^2/2T}{\sqrt{2\pi T}}\left(q + Q + \frac{d}{a} - \frac{2Tc}{a}\right)^{-\bar{m}} dQ \tag{9.12}$$

ith $\bar{m} = -(\lambda + bc/a)$. We leave it up to the reader to note that $\Omega_0(q,T) \to \Omega_0(q)$ as T nds to zero.

To finish let us relate this to some recent work by Feinsilver. Consider (9.10) with $= 0$ and write $\Omega_0(q) = \dfrac{\partial\bar{\Omega}_0(q)}{\partial q}$. Then $\bar{\Omega}_0(q)$ satisfies

$$\left[a\,q\frac{\partial^2}{\partial q^2} + cq\frac{\partial}{\partial q} + m\frac{\partial}{\partial q}\right]\bar{\Omega}_0(q) = 0 \tag{9.3}$$

nd introducing $w_0(q,\gamma) = \exp(-qr)\,\bar{\Omega}_0(q)$, setting

$$B = \frac{a}{ar^2 + cr'}, \quad A = \frac{2ar + c}{ar^2 + cr'}, \quad D = \frac{m}{ar^2 + cr'}, \quad \lambda = -\frac{c+m}{ar+c}$$

ve obtain for $w_0(q,\gamma)$ the equation, $D \equiv \dfrac{\partial}{\partial q}$,

$$[B\,q\,D^2 + A\,q\,D + C\,D + q\,]\,w_0(q,\gamma) = \lambda\,w_0(q,\gamma) \tag{9.4}$$

which has as solution

$$w_0(q,\gamma) = \left[\exp - \frac{(\alpha+\theta)q}{\beta}\;{}_1F_1\left(\frac{\lambda+(\alpha+\theta)\tau}{2\theta}, \tau, \frac{2q\theta}{\beta}\right)\right]$$

where $\alpha = A/2$, $\beta = B$, $\tau = e/B$, $\theta^2 = A^2 - B$ and ${}_rF_s$ is the usual hypergeometric function. This equation plays a central role in Feinsilver's presentation of Lie algebras and recurrence relations. This is not exactly the way I wanted to get it, but here it is anyway.

VI.10 Gauge Transformations and the Cameron-Martin-Girsanov Transformation

In this section we shall examine more closely the effect of the gauge transformations on th processes associated with the Hamiltonians $H_2(p) = p^2/2$ and $H_3(p) = \lambda \int [\exp(p-1)]\mu(d\xi$ considered in section 1, perturbed by a potential $V(q)$.

In each case we shall denote by G_0 the infinitesimal generator of the proces associated to H_2 or H_3 and by G we shall mean $G_0 + V$.

Throughout we shall assume that a positive $\Omega_0(q)$ exists such that $G\,\Omega_0(q) = 0$ an that it is normalized so that it becomes identically 1 when $V(q)$ becomes (tends uniforml to) zero.

We shall compute

$$\tilde{G}\, f(q) = \Omega_0(q)^{-1}\,(G\, f\, \Omega_0)(q)$$

in both cases and verify that

$$\tilde{P}_t\, f(q) = \Omega_0(q)^{-1}\,(P_1\, f\, \Omega_0)(q)$$

is related by the Cameron-Martin-Girsanov transformation to the semigroup of the proces (denoted by $X(t)$ in both cases) with generator G_0 and semigroup $E^q[\, f(X(t))\,]$.

The connection comes from some stochastic calculus applied to the semigroup P which can be written using the Feynman-Kac formula as

$$P_t\, f(q) = E^q\left[\exp \int_0^t V(X(s))\, ds\, f(X(H)\right]$$

as seen in section (V.4).

For $G_0 = \Delta/2$ an easy computation shows that

$$\tilde{G}\, f(q) = \Delta f/2 + \nabla \ln \Omega_0 \circ \nabla f \tag{10.1}$$

which is the infinitesimal generator associated with the Hamiltonian $\tilde{H}(q,p) = (p + \nabla \ln \Omega_0)^2 + V(q)$.

As a side remark we observe that for $V(q) = -(\nabla U(q))^2/2$ for some smooth function $U(q)$, then for $\Omega_0(q) = c \exp U(q)$ we obtain that $\tilde{H}(q,p) = p^2/2 + \nabla U \,.\, P$.

If we denote by $X(t)$ the standard Brownian motion process on $\mathbb{R}^n$, Itô's formula yields

$$f(X(t)) = f(X(0)) + \int_0^t (\nabla f)(X(s)) \,.\, dX(s) + \frac{1}{2}\int_0^t \Delta f(X(s))\, ds$$

or a twice continuously differentiable function $f(q)$.

Since we are assuming that $\Omega_0(q)$ is positive and smooth, it follows from Itô's formula that

$$\frac{\Omega_0(X(t))}{\Omega_0(X(0))} = \exp\left\{ \ln \Omega_0(X(t)) - \ln \Omega_0(X(0)) \right\}$$

$$= \exp\left\{ \int_0^t (\nabla \ln \Omega_0)(X(s)) \,.\, dX(s) + \int_0^t (\Delta \ln \Omega_0)(X(s))\, ds \right\}$$

which can be rewritten, using $\Delta\Omega_0(q)/2 + V(q)\,\Omega_0(q) = 0$, as

$$\frac{\Omega_0(X(t))}{\Omega_0(X(0))} =$$

$$\exp\left\{ \int_0^t (\nabla \ln \Omega_0)(X(s)) \,.\, dX(s) - \frac{1}{2}\int_0^t (\nabla \ln \Omega_0)^2 (X(s))\, ds - \int_0^t V(X(s))\, ds \right\}.$$

It follows from the Feynman-Kac formula and the definition of $\tilde{P}_t$ that

$$\tilde{P}_t f(q) = \Omega_0(q)^{-1} (P_t f\, \Omega_0)(q)$$

$$= E^q\left[f(X(t)) \frac{\Omega_0(X(t))}{\Omega_0(X(0))} \exp \int_0^t V(X(s))ds \right]$$

and using the computation carried out above we obtain

$$\tilde{P}_t f(q) = E^q\left[f(X(t)) \exp\left\{ \int_0^t (\ln \nabla \Omega_0)(X(s)) \,.\, dX(s) - \frac{1}{2}\int_0^t (\nabla \ln \Omega_0)^2 (X(s)) ds \right\} \right]$$

which is the Cameron-Martin-Girsanov representation of the solution to

$$\frac{\partial \tilde{u}}{\partial t} = \tilde{G}\,\tilde{u} = \frac{1}{2}\Delta\tilde{u} + \nabla \ln \Omega_0 \,.\, \nabla u, \quad \tilde{u}(0,q) = f(q)$$

in terms of path integrals over the process $X(t)$.

We shall now repeat the procedure carried above, but this time $X(t)$ will denote th process with infinitesimal generator $G_0 f(q) = \lambda \int (f(\xi + q) - f(q)) \mu(d\xi)$.

In this case $\tilde{G} f(q)$ comes out to be

$$\tilde{G}\, \Omega_0(q) = \lambda\, \Omega_0(q)^{-1} \int (f(q+\xi) - f(q))\, \Omega_0(q+\xi)\, \mu(d\xi). \tag{10.2}$$

We can appreciate the effect of the potential $V(q)$ on the process from (10.2). Th resulting process is again of the "pure jump" type, but the jump rate $\lambda\, \Omega_0(q)^{-1}$ is nov position dependent and the distribution of the jumps is given by $\Omega_0(q+\xi)\, \mu(d\xi)$ depends o the starting point of the jump, i.e. the process is no longer spatcially homogeneous.

Note as well that $\tilde{G}\, 1\, (q) = 0$.

If we again combine the Feynman-Kac formula with the identity

$$\int_0^t V(X(s))\, ds = - \int_0^t \left(\frac{G_0\, \Omega_0}{\Omega_0} \right) (X(s))\, ds$$

we obtain

$$\tilde{P}_t f(q) = E^q \left\{ f(X(t) \frac{\Omega_0(X(t))}{\Omega_0(X(0))} \exp - \int_0^t \left(\frac{G_0\, \Omega_0}{\Omega_0} \right) (X(s)) ds \right\}.$$

We have seen in (V-3.9) that

$$\frac{\Omega_0(X(t))}{\Omega_0(X(0))} \exp - \int_0^t \left(\frac{G_0\, \Omega_0}{\Omega_0} \right) (X(s))\, ds$$

is a Martingale. To bring the analogy with the Cameron-Martin-Girsanov transformation we rewrite it as $M(t)\, N(t)$ where

$$N(t) = \exp \lambda \int_0^t ds \left\{ (G_0 \ln \Omega_0)(X(s)) - \left(\frac{G_0\, \Omega_0}{\Omega_0} \right) (X(s)) \right\} ds$$

and

$$M(t) = \exp \left\{ \sum_{s \leq t} \ln \left(\frac{\Omega_0(X(s))}{\Omega_0(X(s-))} \right) - \lambda \int_0^t ds\, (G_0 \ln \Omega_0)(X(s)) \right\}$$

The exponent in M(t) is a Martingale and it is the analogue of the term

$$\int_0^t (\nabla \ln \Omega_0) \,.\, DX(s).$$

N(t) is what makes the product M(t) N(t) a Martingale

VI.11 From the Invariance Group of the Free Particle to the Invariance Group of the Heath Equation and the Burgers Equation

The outline of this section is the following: we begin by finding the invariance group of the equation

$$\frac{d^2 q}{dt^2} = 0 \tag{11.1}$$

which is Newton's equation of a free particle. In order to verify which of the changes of variable are canonical we have to look at an extended phase space formalism (what better motivation?). After that we extend in an obvious way the representation theory introduced above to obtain the invariance group of the heath equation. Since there is a connection between the heath equation and Burgers equation, we obtain a list of soutions to Burgers equation.

We want to look at al one parameter groups of transformations of the (x,t) plane which leave (11.1) invariant. That is, we want the family

$$\psi_a : \mathbb{R}^2 \to \mathbb{R}^2 \qquad\qquad \psi_a(q,t) = (Q(a,q,t), T(a,q,t))$$

which is smooth in all variables, such that

$$\psi_a(\psi_b(q,t)) = \psi_{a+b}(q,t), \qquad \psi_0(q,t) = (q,t) \tag{11.2}$$

and such that for all a

$$\frac{d^2 Q(a)}{dT^2(a)} = 0 \tag{11.3}$$

which is the meaning of ψ_a being an invariance group for (11.1). The easiest way to get a hold of the group is via its infinitesimal elements. For that, we assume a to be small and write

$$Q(a,x,t) = x + a\,\xi(q,t) + o(a^2)$$

$$T(a,x,t) = t + a\,\tau(q,t) + O(a^2).$$

Computing $dQ(a)/dT$ using the chain rule we obtain

$$\frac{dQ(a)}{dT(a)} = p + a\,\pi(q,t,p) + O(a)$$

where with an eye on what comes below we set $p = dq/dt$ and

$$\pi(q,t,p) = \frac{\partial \xi}{\partial t} + \frac{\partial \xi}{\partial q} - \frac{\partial \tau}{\partial t} p - \frac{\partial t}{\partial q} p^2 . \tag{11.4}$$

if we now require (11.3) to hold, a simple analysis yields

$$\xi(x,t) = c_1 q^2 + c_3 qt + c_6 q + c_7 t + c_8 \tag{11.5-a}$$

$$\tau(\xi,t) = c_3 t + c_2 q + c_1 qt + c_4 t + c_5 \tag{11.5-b}$$

where the c_i's, $i = 1,2,...,8$, are independent constants such that each choice of them determines a group of transformations by solving

$$\frac{dQ(a)}{da} = \xi(Q(a),T(a)) \qquad X(0) = x \tag{11.6-a}$$

$$\frac{dT(a)}{da} = \tau(Q(a),T(a)) \qquad T(0) = t. \tag{11.6-b}$$

Let us list the right simple cases obtained by setting all constants but one equal to zero. The non-zero constant will be given a name consistent with its physical meaning. In this case a has no units. After that, we note that we can pass the units on to a and set the constant equal to 1. Here we go

$$\begin{array}{llll} \text{(i)} & c_5 = t_0, & T(a,q,t) = t + at_0, & Q(a,q,t) = q \\ \text{(ii)} & c_8 = q_0, & T(a,q,t) = t, & Q(a,q,t) = q + aq_0 \\ \text{(iii)} & c_7 = v_0, & T(a,q,t) = t, & Q(a,q,t) = q + av_0 t \\ \text{(iv)} & c_4 = 1, & T(a,q,t) = e^a t, & Q(a,q,t) = q \end{array} \tag{11.7}$$

(iv) $c_4 = 1,$ $T(a,q,t) = e^a t,$ $Q(a,q,t) = q$

(v) $c_0 = 1,$ $T(a,q,t) = t,$ $Q(a,q,t) = e^a q$

(vi) $c_2 = 1/v_0,$ $T(a,q,t) = t + aq/v_0$ $Q(a,q,t) = q$

(vii) $c_1 = 1/L,$ $T(a,q,t) = t/(1 - aw/L),$ $Q(a,q,t) = q/(1 - aq/L)$

(viii) $c_3 = 1/t_0,$ $T(a,q,t) = t/(1 - at_0),$ $Q(a,q,t) = q/(1 - at/t_0)$

Since these transformations involve time, in order to verify which of them is canonical we have to extend the formalism developed in chapter II to treat the time parameter as one more coordinate. This is a well known story. See [II-5] – [II-6].

Instead of (q,p) we now consider (x,t,p,u) coordinates in phase space. We introduce an extended Hamiltonian

$$\mathcal{H}(q,t,p,u) = H(q,p,t) + u$$

and the trajectories in phase space are the curves $(x(s), t(s), p(s), u(s))$ satisfying

$$\dot{q} = \frac{\partial \mathcal{H}}{\partial p} \qquad \dot{p} = -\frac{\partial \mathcal{H}}{\partial q}$$

$$\dot{t} = \frac{\partial \mathcal{H}}{\partial u} \qquad \dot{u} = -\frac{\partial \mathcal{H}}{\partial t}$$

where the dot stands for derivation with respect to s. On account of $\dot{t} = d\mathcal{H}/du \equiv 1$ one can, and does, identity t with s. It is easy to see that the curves now lie in the surfaces $\mathcal{H} =$ constant, and that the motion on each surface is that of the original (time dependent Hamiltonian).

To describe canonical transformations we mimic what we did in chapter II and say that a mapping of the (q,t,p,u) onto the (Q,T,P,U) variables is canonical and generated by a function $G(q,t,P,U)$ whenever the set

$$p = \frac{\partial G}{\partial q} \qquad Q = \frac{\partial G}{\partial P}$$
$$u = \frac{\partial G}{\partial t} \qquad T = \frac{\partial G}{\partial U} \tag{11.8}$$

is such that (Q,T,P,U) can be solved in terms of (q,t,p,u) and vice-versa. When the transformation is known, the set (11.8) can be used to find G, which is the useful object in representation theory.

The other useful criterion for verifying the canonicalness of a coordiante transformation consists in verifying whether or not the following bracket identities hold:

$$[Q,T] = [Q,U] = [T,P] = [P,U] = 0, [X,P] = [U,T] = 1$$

where the extended Poisson brackets are given by (what else?)

$$[f,g] = \frac{\partial f}{\partial q}\frac{\partial g}{\partial p} + \frac{\partial f}{\partial t}\frac{\partial g}{\partial u} - \frac{\partial f}{\partial p}\frac{\partial g}{\partial q} - \frac{\partial f}{\partial u}\frac{\partial g}{\partial t}.$$

In order to complete the list (11.7) we need to add to it the result of integrating

$$\frac{dP(a)}{da} = \pi(Q,T,P) \qquad P(0) = p \tag{11.9}$$

which should be added to the list (11.6). This equation, in whch $\pi(Q,T,P)$ is given by (11.4), can be integrated after the system (11.6) is solved. From the bracket condition

$$[Q(a), P(a)] = 1$$

we obtain, upon differentiation with respect to a and setting $a = 0$,

$$[q,\pi] + [\xi,p] = 0.$$

From this we obtain the identity

$$2\frac{\partial \xi}{\partial q} - \frac{\partial \tau}{\partial t} = 2\,p\,\frac{\partial t}{\partial x}$$

and since the left hand side is independent of p we must have

$$2\frac{\partial \xi}{\partial x} = \frac{\partial \tau}{\partial t}, \qquad \frac{\partial \tau}{\partial x} = 0 \tag{11.10}$$

for the transformation to be canonical. Going back to (11.6) these two conditions imply that $c_1 = c_2 = 0$ and $2\,c_6 = c_4$. This leaves us with only 5 arbitrary constants to play with. Notice as well that the cases (vi) and (vii) drop out of the list (11.7) and, (ii) and (v) are combined into one. So instead of (11.5) we now have

$$\xi(q,t) = c_3\,qt + c_6\,q + c_7\,t + c_8 \qquad \text{(11.5-a)}'$$

$$\tau(q,t) = c_3\,t^2 + 2c_6\,t + c_5. \qquad \text{(11.5-b)}'$$

We can now integrate the set (11.6) and (11.9) to recompile the list (11.7) expanded to

$$
\begin{array}{llll}
\text{(a)} & c_5 = 1: & Q(a) = q, & T(a) = t + a,\ P(a) = p \\
\text{(b)} & c_6 = 1: & Q(a) = p + a, & T(a) = t,\ P(a) = p \\
\text{(c)} & c_7 = 1: & Q(a) = q + at, & T(a) = t,\ P(a) = p + a \\
\text{(d)} & c_4 = 2, c_6 = 1: & Q(a) = e^{a} q, & T(a) = e^{2a} t,\ P(a) = e^{-a} p \\
\text{(e)} & c_3 = 1: & Q(a) = \dfrac{q}{1-at'}, & T(a) = \dfrac{1}{1-at'},\ P(a) = (1-at)p + aq.
\end{array} \tag{11.11}
$$

To find the corresponding $U(a)$ we start from $[T(a), U(a)] = 1$ and notice that this implies, for the cases considered in (11.11) that

$$
\frac{\partial T}{\partial t}\frac{\partial U}{\partial u} = 1 \quad \text{or} \quad U = u\left(\frac{\partial T}{\partial t}\right)^{-1} + f(q,t,p)
$$

where $f(q,t,p)$ is to be determined in each case from $[Q,U] = [P,U] = 0$. After solving these two, we are left with a constant of integration which we shall adjust so that $\tilde{\mathcal{H}}(Q,T,P.U)$ is a multiple of $\mathcal{H}(q,t,p,u) = p^2/2 + u$.

Let us do case (c) explicitly. Substituting $U = u + f(q,t,p)$ in $[Q,U] = 0$ we obtain $\partial f/\partial p + a = 0$ or $f = -ap + f_1(q,t)$ which when substituted in $[P,U] = 0$ yields $f_1 =$ constant which we choose to be $-a^2/2$ so that

$$
\tilde{\mathcal{H}}(Q,T,P,U) = P^2/2 + U = p^2/2 + u.
$$

Had we carried out this procedure for all cases in list (11.11) and then integrated the transformation equations (11.8) to obtain G in each case we would have obtained the following list for G, to which $U(a)$ is added:

$$
\begin{array}{lll}
\text{(a)} & G_a(q,t,P,U) = aP + (t+a)U, & U = u \\
\text{(b)} & G_a(q,t,P,U) = (q+a)P + tU, & U = u \\
\text{(c)} & G_a(q,t,P,U) = (q+at)P + tU - qa - \dfrac{ta^2}{2}, & U = u - pa - a^2/2 \\
\text{(d)} & G_a(q,t,P,U) = e^{a} qP + e^{2a} + U, & U = e^{-2a} u \\
\text{(e)} & G_a(q,t,P,U) = \dfrac{qP}{1-at} + \dfrac{tU}{1-at} - \dfrac{aq^2}{2(1-at)}, & U = (1-at)^2 u - qa(1-at)p - \dfrac{a^2q^2}{2}.
\end{array} \tag{11.12}
$$

In the first three cases $\frac{P^2}{2}+U=\frac{p^2}{2}+u$. In the fourth case $\frac{P^2}{2}+U=\left(\frac{p^2}{2}+u\right)e^{-2a}$ and in the fifth case $\frac{P^2}{2}+U=\frac{\left(\frac{p^2}{2}+u\right)}{(1-at)^2}$. Notice that the surfaces $H=0$ coincide in both cases. This is a restatement of the invariance law given in (11.3).

Notice as well that when we change a to $-a$ and exchange the roles of new and old coordinates we obtain the generating functions for the inverse transformations.

To represent the transformation generated by the $G(a)$ by an operator $T_{G(a)}$ we proceed as in chapter III.

Let us now consider the effect of $T_{\tilde{G}(a)}$ on functions $f(q,t)$. Here $\tilde{G}(a)$ will denote the generating function for going from the $(Q(a), T(a), P(a), U(a))$ to the (q,t,p,u) coordinates. Thus if $f(q,t)$ determines a functional on test functions $\Phi(q,t)$, the theory developed in chapter III yields the following list for $\left(T_{\tilde{G}(a)}\, f\right)(Q,T)$ corresponding to the five cases in (11.12). We have respectively

$$\text{(a)}\quad \tilde{f}(Q,T)=f(Q,T+a)$$

$$\text{(b)}\quad \tilde{f}(Q,T)=f(Q+a,T)$$

$$\text{(c)}\quad \tilde{f}(Q,T)=\exp\left(-Qa-\frac{Ta^2}{2}\right)f(Q+aT,T) \tag{11.13}$$

$$\text{(d)}\quad \tilde{f}(Q,T)=f\left(e^{a}Q,e^{2a}T\right)$$

$$\text{(e)}\quad \tilde{f}(Q,T)=\exp\left(-\frac{aQ^2}{2(1-aT)}\right)f\left(\frac{Q}{1-aT},\frac{T}{1-aT}\right)$$

It is now very easy to verify that if $f(q,t)$ satisfies the equation

$$\frac{\partial f}{\partial t}+\frac{1}{2}\frac{\partial f}{\partial q^2}=0$$

then the $\tilde{f}(Q,T)$ satisfies the corresponding equation in the (Q,T) coordinates, i.e.

$$\frac{\partial \tilde{f}}{\partial T}+\frac{1}{2}\frac{\partial^2 \tilde{f}}{\partial Q^2}=0$$

for each of the five cases in the above list. To understand why one looks at $\frac{\partial f}{\partial t} + \frac{1}{2}\frac{\partial^2 f}{\partial q^2} = 0$ instead of looking at $\frac{\partial f}{\partial t} = \frac{\partial^2 f}{2\partial q^2}$ recall that the latter was given by $f(q,t) = (T_{S(t)}\,\varphi)(q)$ while the former was given by $f(q,t) = (T_{\tilde{S}(t)}\,\varphi)(q)$. Recall the physical meaning of $S(t)$ and $\tilde{S}(t)$: the first one mapped coordinates at t to coordinates at 0 while the second one mapped coordinates at 0 to coordinates at t, which are the ones involved in the mapping $(q,t,p,u) \to (Q,T,P,U)$!

We shall now look at transformations leaving invariant the solution of Burgers equation

$$\frac{\partial W}{\partial t} + W\frac{\partial W}{\partial q} = 0 \qquad W(q,0) = f(q)$$

by analyzing an associated heath equation. Let us introduce $W(\varepsilon)$ satisfying

$$\frac{\partial W}{\partial t} + W(\varepsilon)\,\frac{\partial W(\varepsilon)}{\partial q} + \frac{\varepsilon}{2}\,\frac{\partial^2 W(\varepsilon)}{\partial q^2} = 0 \qquad W(q,0,\varepsilon) = f(q)$$

and introduce ψ by $W = \frac{\partial \psi}{\partial q}$. Substituting and integrating with respect to q we obtain for ψ

$$\frac{\partial \psi}{\partial t} + \frac{1}{2}\left(\frac{\partial \psi}{\partial q}\right)^2 + \frac{\varepsilon}{2}\,\frac{\partial^2 \psi}{\partial q^2} = 0$$

and introducing $\psi = \varepsilon \log \varphi$, φ has to satisfy

$$\frac{\partial \varphi}{\partial t} + \frac{\varepsilon}{2}\,\frac{\partial^2 \varphi}{\partial q} = 0, \qquad \varphi(q,0) = \exp - \frac{1}{\varepsilon}\int_0^q F(q')\,dq'.$$

We know the invariance group of this last equation. The basic one-dimensional subgroups are spelt out in the list (11.13), except the multiplications of φ by a constant which do not change W anyway. Note that in some cases purely imaginary values of a are allowed, like in $a = i\pi$ in (11.13-d). Let us examine the effect of (11.13-c) for example. If we replace $\varphi(q,t)$ by $\tilde{\varphi}(Q,T) = \exp - \left(Qa + \frac{a^2 T}{2}\right)\,\varphi(Q + aT,T)$. Then $\psi(Q,T) = - \varepsilon\left(Qa + \frac{a^2 T}{2}\right) + \varepsilon \log \varphi(Q + aT,T)$ and therefore

$$W_a(Q,T,\varepsilon) = \frac{\partial}{\partial Q}\psi(Q,T) = -\varepsilon a + \varepsilon\frac{\partial}{\partial Q}\log(Q+a,T) = -\varepsilon a + W_0(Q+aT,T,\varepsilon)$$

where W_0 means $\varepsilon\dfrac{\partial \log\varphi(a,t)}{\partial q}$. When ε tends to zero

$$W_a(Q,T,\varepsilon) \to W(Q+aT,T)$$

If we consider (11.13-d) repeating the procedure we would obtain

$$W_a(Q,T,\varepsilon) = e^a\, W_0(e^aQ,e^{2a}T,\varepsilon)$$

which when ε tends to zero yields

$$W_a(Q,T) = e^a\, W(e^a Q,e^{2a}T).$$

And so on and on. Take a look at [IV-11.2] to see that we are doing fine.

The importance of what we did here is the following: if one can go from a flow to a semigroup, then the group of symmetries of the flow becomes a subgroup of the group of symmetries of the semigroup via the representation theory.

References

Sections 1-4 contain variations on the theme of well known results. For comparison check with

[VI-1] Feynman, R. and Hibbs, A. R.: "Quantum mechanics and path integrals". Mc Graw Hill, New York, 1965.

[VI-2] Simon, B.: "Functional integration and quantum physics". Academic Press, New York, 1979.

What put me in the "right track" to reprocess Feinsilver's work is contained in

[VI-3] Gzyl, H.: "Quantum mechanical solution of the linear filtering problem". Letters in systems and control. Vol 3, (1983) 217-220.

And for more on filtering take a look at

[VI-4] Hazewinkel, M. and Willems, J. C.: "Stochastic systems: The mathematics of filtering and applications". D. Reidel Pub, Dordrecht, 1981.

Sections 7 and 8 are based on

[VI-5] Gzyl, H.: "Umbral calculus via integral transforms". Jour. Math. Analysis and Applic. Vol 129, No. 2, (1988) 315-325.

[VI-6] Gzyl, H.: "Evolution semigroups and Hamiltonian flows" Ibid. Vol 110, (1985) 316-332.

For information about the H-W algebra, etc. check with [VI-10] and

[VI-7] Guillemin V. and Sternberg, S.: "Symplectic techniques in physics". Cambridge Univ. Press, Cambridge, 1984.

[VI-8] Feinsilver, P.: "Lie algebras and recurrence relations". Acta Applicandae Mathematicae. Vol 13, (1988) 291-333.

For section 10 check in [VI-2] and [IV-1]. Many books on stochastic processes have such results explained.

For section 11 consult with

[VI-9] Hill, J. M.: "Solution of differential equations by means of one-parameter groups". Pitman Advanced Pub. Program. Boston, 1982.

[VI-10] Benton, E. R. and Platzman: "A table of solutions of the one-dimensional Burgers equation". Quart. Appl. Math. Vol 30, (1972) 195-212.